LEGAL ALCHEMY

◆

LEGAL ALCHEMY

The Use and Misuse of Science in the Law

DAVID L. FAIGMAN

W. H. FREEMAN AND COMPANY ◆ NEW YORK

Text design by Diana Blume

LIBRARY OF CONGRESS CATALOGING-IN-PUBLICATION DATA
Faigman, David L. (David Laurence)
    Legal alchemy : the use and misuse of science in the law / by David L. Faigman.
       p.   cm.
    Includes bibliographical references and index.
    ISBN 0-7167-3143-6
    1. Science and law. 2. Forensic sciences. I. Title.
K487.S3F35   1999
340'.15—dc21   99-31292
                                                     CIP

Printed in the United States of America

First printing, 1999

W. H. Freeman and Company
41 Madison Avenue, New York, NY 10010
Houndmills, Basingstoke RG21 6XS, England

This book is dedicated to three of my college professors—Richard Izzett, Virginia Pratt and Thomas Judd—who together "fixed the destinies" of my life.

Contents

———◆———

PREFACE

◆

Do you believe then that the sciences would ever have arisen and become great if there had not beforehand been magicians, alchemists, astrologers and wizards, who thirsted and hungered after abscondite and forbidden powers?

—FRIEDRICH NIETZSCHE

Weary of science, Dr. Faust turned to magic to summon Mephistopheles. Science could not satisfy Faust's thirst and hunger for knowledge and the power it would convey, for it proved a cumbersome method to achieve those ends. The devil offered a quicker route, though at a greater price. Faust would agree to sell his soul in return for twenty-four years of life during which Mephistopheles would attend to his desires.

The story of the law's use of science bears disturbing similarities to Faust's. The motivations are the same—knowledge and power—and law also displays alarming impatience with the limitations of science's methods. More disturbing still, the law seems all too willing to enter Faustian bargains. Particular outcomes are sought, whatever the route needed to reach them. The boundary between science and magic is also perceived similarly by Faust and by many in the law. They are both tools that might be coopted for service to greater objectives. Efficiency, not verity, is the measure. Very often the legal test lies less in the accuracy of the scientist's or sorcerer's assertions and more in their persuasiveness. Lead can be transmuted into gold, then, as long as people can be convinced of the fact. Whether it really is gold is often beside the point.

The story of Faust was told by both Marlowe and Goethe. The two great writers, however, contemplated very different fates for their protagonist. Marlowe's Dr. Faustus perishes amid the terrible angst he suffers as his part of the bargain becomes due. It is a sorrowful realization after it is already too late. Goethe's Faust, in contrast, is redeemed through service to humanity. In the end, when he falls dead, his soul is borne away by angels before hell can seize it.

The law's fate remains as yet undetermined. That it must choose between science and its lesser cousins is clear. This is the story of law's struggle with the decision; should law work within the complexities and limitations of the subject or seek more immediate rewards but everlasting damnation? As with Faust, the correct choice is obvious. The compelling part of the tale lies in seeing whether law can transcend its base impulses to do the right thing.

For most people, the criminal courts provide the most notable examples of the law's struggle with the promise and peril of science. The popular imagination has been stirred in numerous "cases of the century," almost all of which have involved scientific evidence, from the DNA profile of O. J. Simpson and the drug-induced stupor that "compelled" Richard Allen Davis to kidnap and kill Polly Klaas to the assortment of forensic evidence that was found and lost in the Jon Benet Ramsey investigation. And in a wide variety of less publicized cases, criminal and civil, science is offered to prove a myriad of "syndromes" as well as the toxicity of substances too numerous to count and too scary to contemplate. Rarely does a case, criminal or civil, go to trial today without the presence of experts, most of whom lay claim to a scientific basis for their testimony.

Even brief reflection on the law's use of science soon leads away from the courts to the many policy makers in government who must reconcile the uncertainties of science with the more or less certain convictions they and their constituents hold. In terms of magnitude and management, neither do the courts confront the most scientific information nor are they the worst purveyors of that information. To be sure, the courts today must maneuver through a storm of scientific experts; but this is a warm summer shower compared to the hurricane of science that pours down on administrative agencies. This experience has created agencies that are fairly sophisticated consumers of science—at least when compared to courts. Yet, while courts are accused of being scientifically naive or worse, they are rocket scientists compared to Congress. Although it may be difficult to see, because what emerges from the legislative process seems so scientifically ill informed, legislators come in contact with the largest number and widest array of matters involving science. Instead, politics as usual drives the legislature, and science usually is swept aside, ignored, or corrupted in the process.

Whether it is physician-assisted suicide, silicone breast implants, cloning, desegregation of schools, the reintroduction of the gray wolf into Yellowstone, the standards for ground ozone, or any other of the topics I consider in this book, various government departments are likely to weigh in on the subject—sometimes all at once. "The law" is a skein of interrelated decision makers who have varying perspectives, different constituencies, and divided loyalties. Science thus enters the legal citadel through a variety of doors each having a different doorkeeper responsible for greeting it. These doorkeepers differ in their desire or ability to understand the intricacies of their scientific visitors. Science visits the various wings of the citadel sometimes as an honored guest, sometimes as an unwanted interloper, and sometimes as both at the same time. In most cases, however, science is a poorly understood, not entirely liked or trusted, outsider to the proceedings that ensue. Although the proceedings could not go on without it, there is a score of obstacles that place the scientific profession fundamentally at odds with the legal profession.

I intend to do more, however, than simply present horror stories, real or imagined. Indeed, the story of the relationship between law and science is extraordinarily complex and profoundly subtle. It is not solely about good and evil, though there is much good and sufficient evil to keep the story moving. Instead, it is mainly about miscommunication between two generally well-meaning communities of professionals. A principal reason behind my decision to write this book is to help lawyers and scientists better understand each other.

Lawyers and scientists speak and think in different languages. Imagine the United Nations without translators and you will have an idea of what typical courtroom or legislative and administrative agency hearing rooms are like when scientists testify.

Lawyers and scientists are from different worlds—profoundly different worlds of education and experience. The sorting of professionals into highly compartmentalized categories begins as early as elementary school and is complete by college. Students with an aptitude in math and science gravitate—or are pushed—toward careers in medicine, engineering, physics, chemistry, and the like. Students not so inclined are permitted to avoid real science classes almost entirely or to slip through with "artsy" versions of science courses. Many students who have spent much of their educational life avoiding math and science become lawyers.

Unfortunately, the situation is worse yet. The average lawyer is not merely ignorant of science, he or she has an affirmative aversion to it. Although most lawyers have heard of Stephen Hawking and might have even bought his best-selling book, few, I would venture to guess, have read it, and fewer still have understood it. Indeed, nothing puts a class of law students to sleep faster than putting numbers on the chalkboard. It is a phenomenon you can actually observe. A bell curve makes their eyes glaze over. A minor equation or two or calculating a standard deviation renders law students unconscious; and a more complicated regression analysis induces a deep coma. The average law student's attitude toward mathematics is the same as Huckleberry Finn's: "I had been to school most all the time, and could spell, and read, and write just a little, and could say the multiplication table up to six times seven is thirty-five, and I don't reckon I could ever get any further than that if I was to live forever. I don't take no stock in mathematics, anyway."[1]

More important, law students feel no sense of urgency in learning about science. It is not tested on state bar exams. To the extent that they sense its increasing relevance, most will tell you that experts will assist them in the future. From their vantage point, science is not a necessary part of a legal education.

Most of the fault for this misperception lies with the law schools. Out of the 175 accredited law schools in the United States, only one requires a course in basic statistics or research methodology.[2] Although a large number of schools offer related courses, such as social science in law or psychiatry in law, these are typically taught as seminars and have tiny enrollments. Law students have no sense of urgency about science because their professors do not.

This is an odd failing just about any way you look at it. Law schools typically operate on one of two models of legal education. The so-called elite schools (usually self-defined) offer a glorified liberal arts education that prepares students to be critical-thinking generalists. The so-called nonelite schools (usually defined by others) offer what they consider to be practical and skills-oriented classes that give students the "nuts and bolts" of legal practice. By either standard, science should be a greater part of a law student's education. No lawyer can be a critical-thinking generalist today if he or she cannot distinguish a mean from a mode or an independent variable from a dependent variable. As even a casual review of newspapers and television will support, science permeates every corner of the practice of law.

Without question, law schools will eventually respond to the overwhelming presence of science in the legal process. They have no choice. For now, however, an entire generation of lawyers is being trained without the critical or practical skills to understand what is and what will continue to be an essential part of the lawyer's job description.

The law schools, while at the forefront of much innovative and dynamic thinking, are not on the front lines of the legal process. Judges, legislators, and bureaucrats are this vanguard, and the surge in the importance of science has not gone unnoticed by them. Since 1993, the United States Supreme Court has decided three major scientific evidence cases, *Daubert* v. *Merrell Dow Pharmaceuticals, Inc.* (1993), *General Electric Co.* v. *Joiner* (1997), and *Kumho Tire, Inc.* v. *Carmichael* (1999). These cases (known collectively as "*Daubert* and its progeny") have raised as many questions as they have answered. Moreover, they have placed the matter of science in the law at the top of both federal and state courts' agendas. The House and the Senate, similarly, have both recently considered legislation on the use of science in federal trial courts. And there is not an agency in the country, from the Environmental Protection Agency to Housing and Urban Development, that does not regularly rely on scientific knowledge of one sort or another.

On the front lines, the issue is not the infiltration of science into the legal process but what to do about it. Indeed, ever so slowly, the law is beginning to respond positively, and there are good reasons to be optimistic. But there are also reasons to be concerned. The law's past practice with science has been less than enlightened and often shockingly ignorant. More important, lawyers and scientists continue to see the world through very different lenses. Because neither discipline will or should have to adopt the other's world view, they must reconcile their differences.

In practice, however, scientists are very often frustrated and disgusted by their experience with the law. This is in part attributable to some lawyers' failures both to articulate clearly what their scientific needs are and to comprehend the science and respect the scientists who enter the legal process. Lawyers too are very often frustrated and disgusted by their experience with science. Many scientists fail to make science manageable rather than magical, and some are willing to stretch science to mystical heights for the right fee.

Lawyers need not become scientists, nor scientists lawyers. But since law must rely on science, it is incumbent on lawyers and policy makers to understand it. Most scientists could lead successful and happy lives without ever visiting a courtroom or testifying before a legislative committee or administrative agency. Although lawyers might be happy enough never to meet any scientists, they cannot be successful without them. This is a fact of modern society. As we move into the twenty-first century, the role science plays in our daily lives will continue to increase exponentially. Not only will science be the spectator sport it has been in notorious cases such as O. J. Simpson's, it will play an increasingly essential role in a multitude of diverse issues.

The intersection of law and science, however, is not merely the concern of lawyers, policy makers, judges, and scientists. It is a subject that touches the daily lives of us all. Whether it involves our prurient interest in the forensic evidence of this year's "crime of the century," the toxicity of chemicals in our drinking water, the health effects of eggs, the fate of endangered species, or the exploration of space, it is incumbent on every citizen to appreciate the subject of how lawmakers use science. For the foreseeable future, science will be a necessary and integral part of public policy with which every citizen should struggle. In the end, in a constitutional democracy, the people are responsible for their government's policy. In our technological society, this requires that they too understand how science informs that policy.

ACKNOWLEDGMENTS

◆

A uthors customarily thank numerous people for their assistance but assume the responsibility for any errors that remain. Frankly, I would prefer to blame everyone else for the errors remaining and take any credit that might be due. Unfortunately, since only my friends are likely to read this section of the book, I can be confident that this strategy would lead to disastrous consequences. Hence, at the start, I will grudgingly accept what is inevitably the truth, that the errors remaining are my own (or alternatively the errors of others that I mistakenly allowed them to convince me of). Many, many, people read and commented on sections of the manuscript or sat through endless conversations in which I tried out one idea after another to see if any might fly. Their input contributed immensely to this work. I am deeply indebted to them all.

At the University of California, Hastings College of the Law, I wish to thank my colleagues Jo Carrillo, David Levine, Mary Kay Kane, Evan Lee, Leo Martinez, Roger Park, Eileen Scallen, and especially Ugo Mattei and Radhika Rao for their support, encouragement, and ideas. Outside my home institution, four colleagues merit particular mention: John Monahan, Scott Sundby, Lois Weithorn, and Bill Marshall. John, Scott, and Lois commented on the entire manuscript and more than once saved me from embarrassing errors. Bill, while working in the Office of Legal Counsel in the White House, gave generously of his time and arranged numerous interviews for me with people in various departments of government. Many colleagues around the country also deserve thanks for their ideas and feedback, though they might not always have known I was throwing book ideas at them in our conversations. I want to thank in particular Joe Cecil, Margaret Berger, Ed Imwinkelried, Laird Kirkpatrick, Chris Mueller, Neil Vidmar, Jerry Wetherington, Chris Slobogin, Bob Mosteller, Joe Sanders, Michael Saks, David Kaye, and Charlie Nesson. In addition, many of my close friends suffered through numerous drafts and gave me the moral support and encouragement to keep on writing. Renee Schor and Anna Dalla Val gave me invaluable feedback and particularly warm encouragement. I also want to thank especially Catherine Rogers and Susan Wills for reading the entire manuscript and for demonstrating supreme confidence in me long before it was at all deserved.

Heartfelt thanks and a paragraph all to his own go to Jonathan Cobb. Jonathan was the editor at W. H. Freeman and Company who accepted this manuscript for publication. That alone would merit a thank-you here. But Jonathan left Freeman midway through the writing. He generously offered to continue to read drafts. He also carefully read and edited a full draft of the manuscript this past year, though he was working long hours for another publisher. The final version owes much to his careful eye and ear for the grace of the English language.

At W. H. Freeman, three people merit special thanks. My editors John Michel and Erika Goldman provided invaluable assistance to me. Also, I want to extend my deep gratitude to Sloane Lederer for her early support and constant encouragement. The Freeman team has been delightful to work with.

Finally, I want to thank the Board of Directors of the University of California, Hastings College of the Law. The Board, together with Dean Mary Kay Kane, appointed me as the inaugural holder of the Harry H. and Lillian H. Hastings Research Chair for the year I worked on this book. I hope the dean's and the board's faith in the project is realized, and I am ever so grateful for the generosity of the Hastings family.

LEGAL ALCHEMY

◆

From the Dark Ages to the New Age

The Strange History of Science in the Law

In the beginning God created the heaven and the earth.

And the earth was without form, and void; and darkness was upon the face of the deep. And the Spirit of God moved upon the face of the waters.

And God said, Let there be light: and there was light.

—Genesis

At about one-hundredth of a second after the beginning, . . . the temperature of the universe is 100,000 million degrees Kelvin (10^{11} K). . . . It is filled with an undifferentiated soup of matter and radiation, each particle of which collides very rapidly with the other particles. Thus despite its rapid expansion, the universe is in a state of nearly perfect thermal equilibrium. The contents of the universe are therefore dictated by the rules of statistical mechanics, and do not depend at all on what went before.

—Steven Weinberg, *The First Three Minutes*

On February 11, 1900, a jury returned a verdict of guilty and a sentence of death in the case of the *People of New York* v. *Roland Molineux.*[1] Molineux was charged with the murder of Katherine Adams, who had died as a result of ingesting cyanide, which had been added to a popular headache medication and sent to her nephew through the mails. Her nephew, Harry Cornish, worked at Molineux's athletic club and had recently had a heated dispute with him. The poison, the police surmised, was intended for Cornish. Adams's headache had made his fate hers.

According to the police, this form of mail order poison was Molineux's modus operandi; he had allegedly employed the scheme successfully six weeks before. Early in 1898, Molineux had competed with Henry C. Barnet for the affections of Blanche Cheeseborough. Later reports would indicate that the woman was partial to Barnet. In October 1898, Barnet received a package of Kutnow powders through the mail from an anonymous sender. Shortly after taking some of the powders, he fell desperately ill. Although his physician attributed the illness to diphtheria, Barnet maintained that the cause was "those damned Kutnow powders." It was later determined that the powders contained cyanide of mercury. Barnet died in November. Nineteen days later, Molineux married Blanche Cheeseborough.

Molineux was never charged with the death of Henry Barnet. Nonetheless, the details of Barnet's death and the innuendo surrounding it became a major part of the state's case against Molineux in the trial involving Katherine Adams's death. More damning, however, was the discovery of the private letter boxes rented in the names of H. Cornish and H. C. Barnet. At trial, the person who had let the Barnet box, Nicholas Heckman, would testify—fortified by a promised reward from *The World* (a leader in the yellow journalism of the day)—that Molineux had rented the box from him. Several newspapers soon discovered that these boxes were used to receive sundry pharmaceutical products, especially medications for sexual debility. Molineux denied any connection with these postal boxes.

The core of the state's case, however, came in the form of the "science" of forensic document examination. At trial, the state introduced eighteen witnesses who specialized in handwriting identification to testify that the Cornish and Barnet letters sent to the pharmaceutical companies were written by Molineux. These experts testified further that the writing in the anonymous note that accompanied the package that contained Katherine Adams's death warrant was also in the hand of Roland Molineux. The experts thus were able to provide the empirical link between Molineux and Adams's and Barnet's deaths.

The prosecution's handwriting experts were a motley crew. None had any formal training in handwriting identification. Apparently, American universities did not offer such programs in the late nineteenth century. Most still do not. In addition, none of the prosecution's experts had conducted any research on handwriting patterns in the population or offered data indicating their proficiency in the delegated task. Several were employed as bank tellers and cited their experience and responsibility for evaluating the veracity of signatures for their respective employers. Other than that, these experts had conducted no systematic and rigorous study of handwriting. Fourteen of the handwriting analysts were professional experts, however, with considerable experience testifying in court, though apparently none in checking the accuracy of the conclusions to which they testified. They researched handwriting much as someone might study Milton or Shakespeare. And, as with many "experts" on literature, such musings gave them settled opinions they sought to share with the world. All the state's experts testified confidently to the conclusion that Molineux had penned the questioned documents.

The handwriting experts employed various methodologies to establish the identity of the author of the letters. The bank tellers tended to rely on gestaltlike subjective judgments from "close study" of the questioned writings. As Gilbert B. Sayres, a bank teller for thirteen years, explained, "After I began to make it a study I studied it conscientiously and for a long time, and the more I studied it the more convinced I became of the similarity . . . , the characteristics being the same in both."[2]

The professional experts, on the other hand, employed a more technical and seemingly more sophisticated method of comparison. Persifor Fraser, for instance, a geologist and chemist and a "student of handwriting for about twenty-one years,"[3] testified in somewhat greater detail about the comparison he made between Molineux's handwriting and the note that accompanied the "poison package" received by Cornish. It is worth quoting at some length from Persifor Fraser's testimony. (The numbered exhibits he refers to were known handwriting exemplars provided by the defendant; Exhibit A was the note found with the poison package.)

> Roland Molineux wrote the address on the wrapper, Exhibit A, because, firstly, there are twenty-one characteristics in the conceded writings which are visibly on Exhibit A, secondly, the patterns of a great many letters on the unnumbered [sic] exhibits accord very closely with the patterns of the letters on Exhibit A, and these differences, which exist, are those which would be naturally adapted for disguise; thirdly, the microscopic structure of the ink lines in the numbered exhibits, the conceded writings, agree with the microscopic structure in the ink lines of the Exhibit A in two respects; firstly, in the swelling and tremors, deviations through tremor of the line when highly magnified; secondly, in the characters of the margins of the lines, the edges of the ink lines . . . [and] in searching for some characteristic which was not simply a question of comparison of model or a similarity, it occurred to me that these differences were more than could be accounted for by accident.[4]

There are a number of aspects of the nineteenth-century methods of handwriting identification that seem to be problematic. Most striking, perhaps, is how unscientific the process appears. The experts all knew what results would confirm the hypotheses they were testing. Experimenter bias, usually avoided at all costs in empirical research, was palpable here. The experts also approached the samples looking for confirming instances and were quick to discount or dismiss differences as "adapted for disguise."

The practice of searching a multitude of exemplars for similarities actually turns the scientific method on its head. These experts seemed to take the view that if you have a hundred points of comparison and five constitute "matches," this observation supports the conclusion that the two samples came from the same hand. A less biased method would ask what percentage of matches would be expected if the person did not write the disputed document. A comparison of this number to the number discovered would provide a more accurate statement concerning likely authorship.

The subjective element manifest in the handwriting experts' examinations is tantamount to fraud. Finally, the proficiency of the handwriting experts was never rigorously tested, nor did the courts require tests. In the late nineteenth century, then, although it aspired to scientific status, forensic document examination embraced none of the rigorous methods that would allow us today to label it as a science.

Ultimately, however, what should be most disturbing for the modern reader is the fact that handwriting experts today employ virtually the same methods they used at the turn of the nineteenth century. The lack of empirical validation and failure to conduct proficiency testing, the failure to "blind" testers to expected results and the overwhelming subjective component in the conclusion that the samples "match," are as much a part of handwriting identification analysis today as they were one hundred years ago. In fact, today's handwriting analysts rely on essentially the same authorities as did their nineteenth-century brethren.[5] In the hundred years since Molineux was convicted, largely on the basis of the state's handwriting experts, we have moved from a Newtonian universe to a universe of Einstein, Bohr, and Hawking. We have moved from the biology of Darwin to the DNA helix of Watson and Crick. Cars have replaced the horse and carriage, and planes, spaceships, and satellites fill the sky. And through it all, handwriting experts continue to count similarities and note "swelling and tremors," just as they did a hundred years ago. And courts continue to qualify them as experts.

I do not mean to suggest that handwriting identification cannot be done or that it is not done successfully at times. Common experience suggests that people's handwriting does vary, and there is nothing implausible about developing techniques to identify individual differences. But common experience and lack of implausibility do not make a science. If it did, we would continue to believe that the earth is the center of the universe and that bleeding with leeches is an effective medical therapy. Most aspects of handwriting analysis are subject to tests. But there has been virtually no systematic attempt to study intrawriter variation or interwriter variation, and there have only been sporadic and insufficient attempts to study the reliability of the practitioners of this craft.[6]

An assortment of reasons explain this inertia in the science of handwriting identification analysis. All of these reasons, however, stem from a single cause: market failure. Unlike many other sciences, the primary market for handwriting experts is the law. Neither do they compete among themselves to discover new insights about handwriting comparison, nor do their discoveries have value to other fields. They are a discrete and insular sect of self-validating specialists. They are not trained in the scientific method and they have little clue how to test their claims of expertise. So long as their customers, the courts, keep buying the old model there is no need to come up with anything new. It is as if they began making the Edsel and over the years nondiscriminating car buyers just kept plunking down money for the same old car.

On appeal, Roland Molineux's conviction was overturned largely because the trial court had permitted proof of the defendant's alleged involvement in the Barnet death, a killing with which he was not charged. The appellate court did not object to the surfeit of handwriting testimony. Upon retrial, however, the new judge had little patience for this long queue of handwriting specialists. At one point he exclaimed in exasperation, "What! Another expert?—Well, I suppose we must hear him. Make it quick."[7] In addition, the defense, which had presented no evidence at all in the first trial, responded in the second with its own group of handwriting experts. Not surprisingly, these experts, bought by the defense, testified that the similarities identified by the prosecution were what might be expected in the handwriting of a number of strangers and that the poison package note was not penned by Molineux. This they were confident was true, since the poison package note was written in a better hand than Molineux's. The defendant's expert concluded that one cannot disguise one's handwriting in a better hand than was natural. While this explanation "smells of the lamp" as much as the prosecution's experts' testimony, it apparently had the desired effect on the second Molineux jury. It took the jury just twelve minutes to return a verdict of "not guilty."

Bedrock Principles and Changing Times

As the case of Roland Molineux illustrates, the law does not always stay current with changing science and technology. And this is true at much deeper levels than the courts' failure to insist on validation of handwriting expertise.

Consider a simple thought experiment. Suppose we were able to pluck a Greek citizen from the Athens of Aristotle's time, around 350 B.C. and place him in modern Washington, D.C. That person would probably be able to understand the basic proceedings and decisions made in an assortment of modern legal contexts, from the courtroom to Congress. Certainly, particular mechanisms or procedures have changed, and particular forms of legal relationships have appeared or disappeared, as they have come and gone throughout history. Slavery is gone and democracy is more inclusive than ever. But the rhetoric of politics, the daily business of government, and the resolution of legal disputes have remained much the same through the ages. Our Greek time-traveler, however, would almost surely be completely overwhelmed by our science and technology. Just the ride in the Metro from his hotel to Congress would probably kill him.

This difference in the rates of change between law and science says much about how law and science approach their respective tasks and thus how they inevitably relate to one another. The law's prestige depends largely on adhering to the traditions of the past, while science's prestige turns on how swiftly it advances into the future. But their incompatibility is even more fundamental. Science and law approach the world in profoundly different ways. Even brief reflection reveals stark differences in perspective between the two. Science explores what is; the law dictates what ought to be. Science builds on experience; the law rests on it. Science welcomes innovation, creativity, and challenges to the status quo; the law cherishes the status quo. Science assumes behavior is largely determined by biology and experience; the law typically assumes man has free will.

We might be tempted to conclude that, institutionally, these two professions are so alien to one another that there is little prospect of their ever finding accommodation. In some ways, the personalities of law and science are like those of the tortoise and hare in the familiar Aesop fable. The law's tortoise moves forward deliberately, almost reluctantly, while science's hare bounds forward with great enthusiasm. But in other ways law and science are nothing like the characters in the fable. They are not racing against one another-at least there is no reason for them to be—and it is not even clear that they are going in the same direction. It appears certain, however, that wherever law and science are going, they will not arrive together.

The past can tell us much about these disciplines' respective characters. And more to the point, it can give us insights into their relationship with each other. It turns out that they have more in common than might at first appear to be the case. In fact, in a twist that is reminiscent of a Dickens novel, they are blood relations. The question, though never in doubt in a Dickens story, is whether law and science can live together happily ever after.

Understanding Nature — Including the Nature of Man

Human understanding of the world we inhabit has not always been specifically scientific as we understand the term. To begin with, the word science had a different historical meaning.[8] In ancient times, the word science referred to any body of knowledge that resulted from systematic and rigorous study; it was not limited to the world of experience. Even as late as the nineteenth century, legal scholars presumed that the law itself could be studied as a science. This allowed medieval theologians to claim the scientific mantle.

It was these metaphysicians who produced most of the theories of nature on which the law relied prior to the victory of the modern scientific perspective. The transition from ancient sorcery to modern science, however, was not as smooth or as complete as we might like to believe. Isaac Newton, for instance, not only discovered gravity and charted the heavens using calculus, but he also experimented with alchemy and numerology.[9] Many core insights of astrology remained integral and respected components of science until the late seventeenth century. However reputable science might be today, its roots lie deep in the mystical practices and superstitions of the past.

What we now consider to be within the province of science, previous centuries called the philosophy of nature or natural philosophy. Isaac Newton, it should be recalled, entitled his masterwork *The Mathematical Principles of Natural Philosophy*. Although Newton understood his task as part of a broader philosophical investigation of why the world takes the shape humans confront, he shared the specific goal of contemporary scientists of describing how the world works. The move from a focus on the how and the why of nature to solely on the how is generally associated with the scientific revolution. The scientific revolution ushered in the modern view that science may study what can be tested and leave what cannot to priests, philosophers, and sorcerers.

There is great power in having the capacity to describe how our world works. But there is greater power still in knowing the reasons why it does so and, even more, in dictating how it should do so. The objectives of the science of today and the natural philosophy and religion or mysticism of the past are nearly identical. Man has consistently sought knowledge of himself and the world that surrounds him in order to better predict and control the chaos of daily living. There is no fundamental difference between ancient man's attempt to predict and control the floods of the Nile and modern man's attempt to do the same.[10] Only the methods used have changed over time. Modern man is more likely to rely on his understanding of the applied physics of meteorology and principles of engineering than astrology or ancient ritual. But the mystical practices of the past have hardly been abandoned. While modern engineers are employing the latest technology against flooding, invariably their efforts are accompanied by ritual and prayer from many quarters.

Law also has deeply religious roots, and its similarities to religious practice are plain. Religion has proved to be a highly effective device for setting and controlling the boundaries of human behavior. The law shares these objectives, so it should not be surprising to find that lawmakers also use the methods of religion. Whether explicit or implicit, lawyers and policy makers regularly appeal to the bedrock moral principles from time immemorial. But this is not mere mimicry. In most societies religion either was the law or largely controlled it. Even today, around the world this is more likely to be true than not. American law closely resembles the religious kingdom because it traces its roots there.

For most of human history law and science were largely unified within the corpus of religion. It is for good reason, then, that lawyers and scientists are sometimes individually referred to as "the high priests" of modern society. Historically, the real high priests were a combination of "lawyer" (or judge), "scientist," and "priest." So long as church and state were one, no conflict arose. Religion described the natural order of the universe and explained the moral dictates that followed naturally from that reality. Religion supplied both a description of the natural world and the punishment for failing to conform to the rules mandated by that world.

Science eventually separated from religion and, in time, began to challenge the authority of religion to describe the natural world. But the connection between law and science cannot be understood without appreciating their common heritage in religious faith and practice. This heritage remains highly relevant to much of their modern relationship. Moreover, as interesting as this common ancestry is, it supplies only a part of the story. The other part must be sought in the development of law, science, and religion after they separated from one another.

That law, science, and religion started life together is not terribly profound in itself. What has occurred since they separated into different institutions is of greater relevance here. Each retained to some degree an interest and claim to expertise in the specialty of the other two. Religion, of course, never surrendered its claim of authority to describe the empirical world and prescribe what was necessary to move on to a better world. Of much greater significance, both law and science retained or adopted religious overtones that reflected their common roots with religion. In the United States, at least, the law has developed a complex personality that unites facets of religious belief and superstition with the faith that the scientific enterprise will provide deliverance. In effect, American law has itself become a sort of religious institution, but one that must account for secular scientific findings. Science too has developed a rich sense of religious purpose, though not all scientists share it equally. In addition, scientists have increasingly sought a voice in lawmaking and policy formation. In a nutshell, whereas, in the past, religion dominated the field, the division of law, science and religion into separate institutions has led to a competition among the three for the hearts, minds and souls of society.

Many observers have likened the contemporary relationship between law and science to that of a romantic courtship. As one commentator suggested, the "meeting grounds" of law and science "are rather like the parlor in the Victorian home in which the girl and her suitor can get together—but not get together too much."[11] But law and science have a much longer and significantly more complex history than the courtship metaphor can capture.

I prefer a more dramatic metaphor. In Greek mythology, the Hydra was a many-headed beast that lived in the swamps of Lerna. As penance for a crime, Heracles was sentenced to a series of heroic tasks, including slaying the Hydra. This was no easy task, since the Hydra grew two heads for every one that Heracles lopped off. Heracles eventually prevailed in his battle with the Hydra, with the aid of his twin brother's son, Iolaus, who used a torch to cauterize the wounds so that new heads could not grow as his uncle lopped them off. Imagine now a three-headed hydra, with the heads representing law, science, and religion. Each lopped-off head would become its own three-headed hydra. Any hero wishing to fight the three separate three-headed Hydras currently before us will have a far greater task than Heracles faced. The trick will be to lop off two of the heads of each of the three beasts and then to make these now one-headed creatures live in harmony.

The Science Hydra

When Science Makes Law

From Einstein's famous letter to Franklin Roosevelt about nuclear fission to scientists' experimentation with transplanting pig hearts into humans ("xenotransplantation"), science challenges society to react. This is a natural consequence of one field developing techniques or ideas that are of pressing importance to another. Problems arise, however, when scientists go beyond contributing to policy agendas and begin setting policy. It is a lot like the division of authority between the military and civilian sectors of society. When the Douglas MacArthurs of the scientific community start insisting on "victory at any cost," it is time for civilians to reassert control. History is replete with examples of both good scientists and bad scientists urging what they considered to be "inevitable" policy choices based on their science. In reality, science rarely, if ever, compels a particular policy choice. And even if it does, it still falls to the civilians to recognize that fact.

A particularly salient—and sad—example of scientists stepping into policy making comes from several of the more recent volumes in the library of human prejudice. According to evolutionary theorists, such as Steven Pinker and Edward O. Wilson, the modern tendency to define in-groups and out-groups is a remnant of our hunter-gatherer pasts. Hence, bigotry, hypernationalism, religious intolerance and maybe even the movie *Independence Day* can all be traced to the evolutionary advantages tribalism played among our distant ancestors. It is ironic, then, that since Darwin first wrote about the subject, many have employed evolution to justify these prejudices. Scientists play an unfortunately prominent role in the story, though admittedly there is plenty of blame to go around. Francis Galton's eugenic ideas illustrate the all-too-slippery slope between what scientific research appears to suggest and what scientists claim their research dictates.

Francis Galton was Darwin's cousin and an accomplished mathematician who contributed many significant ideas to the development of statistical theory. He was also a thoroughly dangerous thinker. He first proposed his eugenics ideas in 1865, six years after *Origin of Species* first appeared. Darwin's *Origin* produced an epiphany for Galton, one that he devoted much of his life both to studying and to advocating. He told his cousin, "Your book drove away the constraint of my old superstition, as if it had been a nightmare."[12] Eugenics would allow man to control his destiny, breeding supermen for a glorious future society. Galton explained, "What Nature does blindly, slowly, and ruthlessly, man may do providently, quickly, and kindly."[13] Eugenics provided a scientific gloss to a very old emotion.

In the twentieth century, eugenics was supported by many scientists and policy makers who hoped to breed "superior people." Galton tended toward this "positive" form of eugenics, believing that society could be improved by encouraging "quality stock" to breed together. But eugenics has a dark side that cannot be wholly separated from even the most positive spin. Galton himself regularly slipped over to this dark side when he advocated the comfortable segregation of the unworthy to monasteries and convents.[14] In time, this so-called positive eugenics was joined by many adherents who openly preached that "undesirable citizens . . . must not be bred."[15] It was this negative perspective that would come to dominate the debate, as mainstream eugenicists increasingly sought effective ways to stop the flood of degenerates from washing over the continent. Karl Pearson, another famous statistician, headed the laboratory that Galton, his mentor, had established. Pearson insisted that his laboratory had no political agenda: "We of the Galton Laboratory have no axes to grind," he declared. "We gain nothing, we lose nothing, by the establishment of the truth."[16] The laboratory's attitude was summarized by Ethel M. Elderton: "Improvement in social conditions will not compensate for a bad hereditary influence. . . . The only way to keep a nation strong mentally and physically is to see to it that each new generation is derived chiefly from the fitter members of the generation before."[17] Dr. William J. Robinson summarized this sentiment more bluntly: "It is the acme of stupidity to talk in such cases of individual liberty, of the rights of the individual. Such individuals have no rights. They have no right in the first instance to be born, but having been born, they have no right to propagate their kind."[18]

Individual liberty, however, is the stuff of the law. It is as much aspiration as it is definition. Although science might one day "prove" that all people are not created equal, the law need never accept such "truth," for legal principle can never be simply about base genetic reality. The law certainly must be grounded in the best science has to offer. But the law is also about being ungrounded and in aspiring toward higher ideals. Thomas Jefferson should be our model. However much we smug members of modern society might condemn him for his shortcomings in this very area of prejudice, he understood that the law sought a higher plane. No amount of science can counter those truths he knew to be self-evident, "that all men are created equal; that they are endowed by their creator with certain unalienable rights; that among these are life, liberty, and the pursuit of happiness."[19] These truths cannot be falsified by science; they are self-evident matters of policy.

Looking for God Through Science

The next great task of science is to create a religion for mankind.
—LORD JOHN MORLEY [of Blackburn]

Given the long history of speculation about the empirical universe—born out of religion, spiritualism, abstract reasoning, fear, and loathing—modern scientists cannot hope to completely avoid conflict with the priests, philosophers, and sorcerers who continue to ply their respective trades. In describing how the world works, science necessarily finds itself in conflict with older natural theories that described how the world works from premises that incorporated why the world is the way it is or why it should be some other way. By contradicting previously held natural philosophies, scientists inevitably cast doubt on the religious or moral systems that are based on them. In Genesis, God's creation of the universe portends significant moral lessons that soon follow the dramatic opening scene. For a physicist like Steven Weinberg, the description of the universe's first moments entails no similar profound message. In the same way, although Darwin mainly avoided any theological statement in his description of evolution—at least until his final paragraph—his explanation of how humans came to be human avowedly contradicted the Old Testament's version. Darwin was intimately familiar with the ramifications of his research, just as Newton and Galileo before him understood the consequences of their work. Indeed, Darwin delayed publishing the *Origin of Species* for decades and suffered debilitating illness as a consequence of the anxiety that publishing his theory of natural selection caused him.[20] Science thus shares a tradition with all heresies. By questioning preexisting doctrine, it challenges not only the nature of the world but also, inevitably, its meaning.

The association between God (read morals) and the empirical world is central to appreciating the modern relationship between law and science. In the past, lawgivers typically drew a close connection between the empirical world and the moral universe. Because these lawgivers were usually religiously affiliated, a challenge to the religious understanding of the way the world is was tantamount to a challenge of the moral order. This made the early scientists heretics, even if they sought only to describe the factual nature of the universe. They were dangerous because their science inevitably also challenged the moral state of the universe. The scientific revolution initiated a basic change by separating the is from the ought. But old habits die hard. We continue to reason from what is natural to what is moral, though this connection has become considerably more complicated in modern times. In the past, however, this connection was relatively straightforward and challenges to either component of it were rigorously resisted. As we will see, matters have not changed considerably over time.

Even the most sophisticated modern scientists have not entirely forsaken the metaphysical questions that lurk behind the hypotheses designed to answer how the world works. It is seemingly an inherent part of the human character to look for the objective behind the object. Many scientists cannot resist going beyond how things work to speculate about why they do so. When this occurs, as Hanns Johst said in describing his reaction to hearing the word "culture," it is usually time to reach for your revolver.[21]

It has become increasingly commonplace, for instance, for physicists to inject the deity into, at least, their popular science writing. Whether it is Leon Lederman describing the "God Particle," Paul Davies finding common ground between "God and the New Physics," or Stephen Hawking searching for the mind of God in black holes or other cosmological exotica, the need to identify the purpose of it all appears universally human. Hawking described the relationship between knowing God and knowing physics in the conclusion to his best-selling book, *A Brief History of Time*:

> If we do discover a complete [unified] theory [of the universe], it should in time be understandable in broad principle by everyone, not just a few scientists. Then we shall all, philosophers, scientists, and just ordinary people, be able to take part in the discussion of the question of why it is that we and the universe exist. If we find the answer to that, it would be the ultimate triumph of human reason—for then we should know the mind of God.[22]

Hawking's presumptuousness likely was born out of the grandeur of the admittedly magnificent feat he described of deriving a grand unified theory. But it is far from obvious that knowing how the universe was created will tell us why it was created. Also, we might inquire, why won't other major scientific discoveries, such as mapping the human genome, or other technological feats, such as cloning, not show us the mind of God? Indeed, we should ask, why doesn't the simplest observation of our universe give us this insight? Can we not know the mind of God when we look at a flower, see the flight of a butterfly, or hear the sound of the surf? As William Blake urged:

> To see a world in a grain of sand
> And a heaven in a wild flower,
> Hold infinity in the palm of your hand
> And eternity in an hour.[23]

In the realm of faith and religion, Hawking should have hesitated over Pascal's admonition: "It is the heart which experiences God, and not the reason."[24]

Scientists' predilections to speculate about religion and to advocate the "good society" are related. They seek to broaden their field of influence from saying what is to also speculating on how it came to be and how it ought to be. The error scientists make is one of arrogance. And it is an error as old as the scientific revolution. While what *is* might very well assist us to decide what *ought to be*, it never dictates what *must be*. In confusing the is with the ought, scientists make the same mistake, albeit in reverse, that priests, philosophers, and sorcerers have made throughout history. Whereas a scientist speculates about metaphysics from how the real world is situated, a metaphysician speculates about the real world in order to situate his moral values. In either case, as they reach beyond their fields of expertise, conflict results.

The Religion Hydra

Knowing Reality Through Faith

The first truly profound thing our Athenian citizen from 350 B.C. would learn about modern Washington is that it and the earth are no longer the center of the universe (even if this fact might surprise some modern-day Washingtonians). But the realization that the earth was not at the center would be profoundly disturbing to him in a way that can scarcely be imagined today. The validity of the geocentric theory of the universe was the subject of one of the best-known and most celebrated "trials" in history, a trial that pitted the new science against the old theology. The inherent conflict between science and modern values that continues today was starkly presented in the greatest trial of the fifteenth century: *The Church v. Galileo Galilei.*

In embracing, revising, and providing substantial evidence for Copernicus's heliocentric theory, Galileo challenged the Aristotelian world view that conformed so comfortably with both experience and the Bible. By any measure, Aristotle was one of the greatest natural philosophers in all history. Aristotle's description of the universe was born of the stuff of modern science—observation. To be sure, Aristotle did not share our modern sensibilities, especially regarding the experimental method, but he was a close observer of natural phenomena.[25] He was a scientist in the tradition Thomas Huxley described when he commented that "science is nothing but trained and organized common sense."[26] As we can certainly appreciate today, it is obvious to the senses that the sun revolves around the earth. Aristotle's genius and "organized common sense" similarly supported this conclusion:

> This view is further supported by the contributions of mathematicians to astronomy, since the observations made as the shapes change by which the

order of the stars is determined, are fully accounted for on the hypothesis that the earth is at the center.[27]

The Bible raises this common observation to an article of faith. The psalmist proclaims straightforwardly that God "fixed the earth upon its foundation, not to be moved forever." (Psalms 103:5). The Book of Ecclesiastes observes that "the sun rises and the sun goes down: then it presses on to the place where it rises." When Galileo openly challenged the geocentric theory, therefore, much turned on his claim. How, people wondered, should the text of Josue (Joshua 10:12–13) be interpreted? "Josue prayed to the Lord, and said in the presence of Israel, 'Stand still, O sun, at Gabaon, O moon, in the valley of Aialon!' And the sun stood still, and the moon stayed, while the nation took vengeance on its foes." If Galileo was correct, of course, there was no need to command the sun to stand still—it already obeyed this admonition. Both the value of our ordinary senses as well as the integrity of our faith were thus challenged by Galileo's science. It is no wonder that the "law" sought to crush this heresy.

In the fifteenth century, it was simply known beyond serious question that the earth was the center of the universe. Controversy over the stellar configuration mainly revolved around whether Aristotle's (378–322 B.C.) or Ptolemy's (ca. 150 B.C.) model better reconciled the Bible with daily experience. Both models were geocentric and each required a substantial number of artificial assumptions to make them work. During Aristotle's time and to a lesser extent Ptolemy's, as well, natural philosophers did not aspire to exact quantitative agreement between theory and observations. The astronomy of the fourth century B.C. in this way resembled much of psychological science of the twentieth century. Rough approximation was considered a high achievement. It was this lack of agreement between the models and reality that led Pope Leo X in 1514 to call on experts in theology and astronomy for corrections that would enable the church to reform the ecclesiastical calendar. Nicolaus Copernicus answered this call.

Conveniently for my story, Copernicus's principal area of study was the law. He studied church law at the University of Bologna, one of the finest European law schools, and graduated with an advanced degree in canon law from the University of Padua around 1503.[28] Together with the law, Copernicus studied medicine, theology, and, of course, mathematics and astronomy. Although he was not a priest, Copernicus spent most of his working years at the Cathedral of Frauenberg, where he conducted his duties as a canon. Throughout these years he sought an answer to the "lack of certitude in the traditional mathematics concerning the composition of movements of the spheres of the world."[29] His answer literally turned the world upside down. Fearing embarrassment more than persecution, Copernicus waited until 1530 to circulate an outline of his new astronomy. He had good reason to be fearful. For instance, in 1533, Martin Luther condemned his theory with scorn:

> People give ear to an upstart astrologer who strove to show that the earth revolves, not the heavens or the firmament, the sun and the moon. Whosoever wishes to appear clever must devise some new system which of all systems, of course, is the very best. This fool wishes to reverse the entire science of astronomy; but Sacred Scripture tells us that Josue commanded the sun to stand still, and not the earth.[30]

Copernicus's great work was not published until 1543. He received an advance copy on May 24 of that year as he lay on his deathbed. He died a few hours later. The work, *De revolutionibus orbium coelestium*, was dedicated to Pope Paul III.

If Copernicus's theory cannot be said to have received an overly warm reception in Rome, it was not condemned or banned—at least not yet. This cordial reception was due, at least in part, to the unsigned preface that accompanied the work, written by Andreas Osiander. Osiander, a Lutheran theologian, seeking to avoid theological complications, asserted that the heliocentric theory was merely an hypothesis that was useful for computing the calendar, not a representation of reality. The preface, moreover, was written in a way that suggested that Copernicus himself had composed it. It is quite certain, however, that Copernicus believed that his theory described reality. But, in the church's view, so long as it remained only a hypothesis, it was not threatening enough to ban.

Although in time the church would ban the teaching of Copernicus and censure and eventually condemn Galileo for heresy, its position did not change substantially over the years. The church, it is true, stood steadfastly by its literal interpretation of the Bible, concluding that the theory that the earth revolves around the sun was heretical. The church, however, generally drew a distinction between hypotheses and fact and was inclined to be somewhat tolerant of the heliocentric heresy so long as it was not put forth as fact and thus contrary to the Bible. For Galileo, problems arose when he insisted that scripture should bow before his proof and be interpreted in light of the facts as he found them to be.

The trial of Galileo is almost as well known today as some of the famous trials of recent years. And like its twentieth-century counterparts, Galileo's trial was filled with intrigue and much ambiguity continues to shroud the basic facts. The essential story and its object lessons can be quickly summarized. Galileo was tried for heresy and for allegedly violating an injunction served on him in 1616 "not to hold, teach, or defend in any way, verbally or in writing" the Copernican hypothesis. The specific target of the trial was Galileo's monumental work *Dialogue on the Great World Systems*. The *Dialogue* presented a Socratic colloquy between Salviati, who represented the new learning, and Simplicio, who defended the old. The moderator between the past and the future was Sagredo. Although Galileo may have intended the *Dialogue* to be a balanced presentation, as he maintained at his trial in 1633, it was understood immediately as a devastating critique of Aristotelian philosophy and Ptolemaic astronomy.

Despite his lack of subtlety in the *Dialogue*, Galileo's defense was that he had never held the view that the earth revolved around the sun. In response to interrogation, "with the threat of torture,"[31] Galileo stated that from the time of the 1616 injunction he had held "as most true and indisputable, the opinion of Ptolemy, that is to say, the stability of the earth and the motion of the sun."[32] Moreover, despite the apparent arguments forwarded by the Dialogue, Galileo stated categorically, "I affirm therefore, on my conscience, that I do not now hold the condemned opinion and have not held it since the decision of the authorities."[33]

While physical torture was never really a serious threat, both the process and its conclusion must have seemed to Galileo torture enough. The court convicted Galileo of the heresy of believing "that the sun is the center of the world and does not move from east to west and that the earth moves and is not the center of the world."[34] Galileo was sentenced to "abjure, curse, and detest" his errors and heresies and to life imprisonment, which he served under house arrest in his villa outside Florence. In addition, the *Dialogue* was "prohibited by public edict."

Although Galileo's condemnation was now complete, the church's would soon begin. The church had wielded raw power to quash the new physics. As the proverbial saying goes, it won the battle but was destined to lose the war. Despite the fact that Copernicus had broached the theory more than sixty years before, the church and society generally were not prepared for the new science. It wished that the new learning would go away, and it had the worldly power to make it go away. But the church could make it disappear only for a limited time. When it returned and prevailed, the church would lose more in credibility and prestige than it ever stood to gain by its appeal to faith and its condemnation of the unfaithful. There is a valuable lesson for all lawmakers in this tale.

The church's reaction to Galileo also represents a juncture of sorts in Western thinking. The need to divorce science and faith was becoming increasingly apparent, though it would take over three hundred years for this doctrine to become firmly entrenched. As science began to tread on the domain previously controlled by priests and sorcerers, these "faiths" had only two choices. They could compete head to head with science or they could strategically withdraw to another plane from which they could make claims to knowledge separate from science. In the West, and especially in the United States, faith retreated to fight another day. The separation of church and state is one manifestation of this withdrawal.

In fact, one aspect of the Galileo affair anticipates a strategy that is central to the tactics employed by the faithful in the modern conflict between church and state. The church was grudgingly tolerant of the heliocentric theory so long as it was proffered merely as a hypothesis that offered advances in computational accuracy, rather than a description of reality. The church was willing to share the intellectual stage with science, so long as science understood who was the star. Today, of course, the church no longer has the star power it enjoyed in the sixteenth century. The modern church shares the lament of Norma Desmond in the movie *Sunset Boulevard:*

> JOE GILLIS: You used to be in pictures. You used to be big.
> NORMA DESMOND: I am big. It's the pictures that got small.[35]

Making Law Through Religion

> *T*hus, from the war of nature, from famine and death, the most exalted object which we are capable of conceiving, namely, the production of the higher animals, directly follows. There is grandeur in this view of life, with its several powers, having been originally breathed by the Creator into a few forms or into one; and that, whilst this planet has gone cycling on according to the fixed law of gravity, from so simple a beginning endless forms most beautiful and most wonderful have been, and are being evolved.
>
> —CHARLES DARWIN,
> *Origin of Species* (final paragraph)

The United States has been called the most religious nation on earth. Although this might be a generous description, it bespeaks at least the strong position religion occupies in American society. Yet we are also a land in which the state is constitutionally separated from the church. This might appear to be a paradox of sorts, since a nation devoted to religion would presumably wish the state to back up that devotion. The

Founding Fathers, however, saw no paradox, for they believed that the state must be kept separate from the church for religion's own good. Somewhat contrary to the modern fear that religion will usurp the state's functions, the framers of the Constitution feared that the state would usurp religion. The First Amendment contains two clauses directed at this matter. One guarantees the individual the right to freely exercise his or her religion and the other forbids the state from "establishing" religion. In short, government is forbidden from interfering with both religious individuals and religious institutions.

In contrast, religions are not prohibited from fully participating in the affairs of government. The Supreme Court has held repeatedly that religions must be allowed to contribute ideas and influence just as other institutions do in a participatory democracy. Still, government cannot adopt policies in the name of religion. Hence, Pat Robinson, an ordained minister, could run for president in the name of religion, and he could advocate policies that his followers believe in, but he could not govern on the basis of religious premises. He would need to articulate an alternative argument to support his action in the unlikely event that he should become President.

The two religion clauses thus require a balancing act of death-defying proportions. Walking the tightrope between the free exercise and establishment clauses creates a fascinating dynamic between the church and the state. Most religions continue to have strong prescriptive programs for society. But religion in the United States does not have the power of the sixteenth-century Vatican, so it must seek out more indirect ways to satisfy its lawmaking desires. The courts' (and the Constitution's) task is to ferret out policies driven by religious zeal and to permit policies with independent secular merit. Just as in the sixteenth century, the church continues to be informed by its view of nature. As part of its regulatory agenda, the church often tries to impose this view on secular society.

By the twentieth century, of course, Copernicus's heliocentric theory had fully won out. The church has for some time reconciled itself to a mobile earth and a fixed sun. In fact, in a 1994 memorandum, Pope John Paul II apologized for the church's past sins against Galileo. Now that science was the star, the new question was to what extent faith might share the intellectual stage. By the late nineteenth century and throughout the twentieth, evolution emerged as the issue that would define the borderland where law, science, and faith met. How much as well as how little things had changed can be seen in the 1925 prosecution of John Scopes.[36] The so-called Scopes Monkey trial, which was profoundly symbolic but had little legal significance, was a circus to match any that came before or would follow.

John Scopes, a high school biology teacher, had challenged a Tennessee law that prohibited the teaching of evolution. The law made it unlawful "to teach any theory that denies the story of divine creation as taught by the Bible and to teach instead that man was descended from a lower order of animals." Scopes was represented by a team of lawyers that included the irascible Clarence Darrow. The case for Tennessee was led by the pious William Jennings Bryan, a populist and three-time Democratic candidate for president. Part of the trial was held on the courthouse lawn with a crowd of more than 5,000 in attendance. All of the proceedings were brought to a national audience by an army of reporters, including the inimitable H. L. Mencken. Around the courthouse, banners flew and lemonade stands offered relief from the July heat. Chimpanzees, said to be witnesses for the prosecution, performed in a sideshow on Main street. The Scopes trial was an early illustration of the spectacle that celebrity trials could become in the modern world.

The Scopes trial was never really about whether John Scopes was guilty, since he had deliberately violated the law in order to test its constitutionality. But the trial also was not really about the law's constitutionality, since this determination is not triable by a jury—it is decided by a court as a matter of law. In the end, like so many other modern trial spectacles, the trial was about American society and what happens when values clash. No better illustration of the clash between sectarian piety and secular zeal could be offered than the courtroom confrontation between William Jennings Bryan and Clarence Darrow. In a highly unusual procedural move, Darrow called Bryan to the stand and cross-examined him on interpreting the Bible. The following famous exchange occurred on the seventh day of the trial:

> DARROW: You have given considerable study to the Bible, haven't you, Mr. Bryan?
> BRYAN: Yes, sir, I have tried to.
> DARROW: Then you have made a general study of it?
> BRYAN: Yes, I have; I have studied the Bible for about fifty years, or some time more than that, but, of course, I have studied it more as I have become older than when I was a boy.
> DARROW: You claim that everything in the Bible should be literally interpreted?
> BRYAN: I believe everything in the Bible should be accepted as it is given there: some of the Bible is given illustratively. For instance: "Ye are the salt of the earth." I would not insist that man was actually salt, or that he had flesh of salt, but it is used in the sense of salt as saying God's people.

DARROW: But when you read that Jonah swallowed the whale—or that the whale swallowed Jonah—excuse me please—how do you literally interpret that?

BRYAN: When I read that a big fish swallowed Jonah—it does not say whale. That is my recollection of it. A big fish, and I believe it, and I believe in a God who can make a whale and can make a man and make both what He pleases.

DARROW: Now, you say, the big fish swallowed Jonah, and he there remained how long—three days—and then he spewed him upon the land. You believe that the big fish was made to swallow Jonah?

BRYAN: I am not prepared to say that; the Bible merely says it was done.

DARROW: You don't know whether it was the ordinary run of fish, or made for that purpose?

BRYAN: You may guess; you evolutionists guess. . . .

DARROW: You are not prepared to say whether that fish was made especially to swallow a man or not?

BRYAN: The Bible doesn't say, so I am not prepared to say.

DARROW: But do you believe He made them—that He made such a fish and that it was big enough to swallow Jonah?

BRYAN: Yes, sir. Let me add: One miracle is just as easy to believe as another.

DARROW: Just as hard?

BRYAN: It is hard to believe for you, but easy for me. A miracle is a thing performed beyond what man can perform. When you get within the realm of miracles; and it is just as easy to believe the miracle of Jonah as any other miracle in the Bible.

DARROW: Perfectly easy to believe that Jonah swallowed the whale?

BRYAN: If the Bible said so; the Bible doesn't make as extreme statements as evolutionists do. . . .

DARROW: The Bible says Joshua commanded the sun to stand still for the purpose of lengthening the day, doesn't it, and you believe it?

Bryan: I do.

DARROW: Do you believe at that time the entire sun went around the earth?

BRYAN: No, I believe that the earth goes around the sun.

DARROW: Do you believe that the men who wrote it thought that the day could be lengthened or that the sun could be stopped?

BRYAN: I don't know what they thought.

DARROW: You don't know?

BRYAN: I think they wrote the fact without expressing their own thoughts.

DARROW: If the day was lengthened by stopping either the earth or the sun, it must have been the earth?

BRYAN: Well, I should say so.

DARROW: Now, Mr. Bryan, have you ever pondered what would have happened to the earth if it had stood still?

BRYAN: No.

DARROW: You have not?

BRYAN: No; the God I believe in could have taken care of that, Mr. Darrow.

DARROW: I see. Have you ever pondered what would naturally happen to the earth if it stood still suddenly?

BRYAN: No.

DARROW: Don't you know it would have been converted into molten mass of matter?

BRYAN: You testify to that when you get on the stand, I will give you a chance.

DARROW: Don't you believe it?

BRYAN: I would want to hear expert testimony on that.

DARROW: You have never investigated that subject?

BRYAN: I don't think I have ever had the question asked.

DARROW: Or ever thought of it?

BRYAN: I have been too busy on things that I thought were of more importance than that.

DARROW: You believe the story of the flood to be a literal interpretation?

BRYAN: Yes, sir.

DARROW: When was that flood?

BRYAN: I would not attempt to fix the date. The date is fixed, as suggested this morning.

DARROW: About 4004 B.C.?

BRYAN: That has been the estimate of a man that is accepted today. I would not say it is accurate.

DARROW: That estimate is printed in the Bible?

BRYAN: Everybody knows, at least, I think most of the people know, that was the estimate given.

DARROW: But what do you think the Bible itself says? Don't you know how it was arrived at?

BRYAN: I never made a calculation.

DARROW: A calculation from what?

BRYAN: I could not say.

DARROW: From the generations of man?

BRYAN: I would not want to say that.

DARROW: What do you think?

BRYAN: I do not think about things I don't think about.

DARROW: Do you think about things you do think about?

BRYAN: Well, sometimes.[37]

In a nation in which church and state are ostensibly separate, outright banning of scientific theory was not likely to persist. And it did not. But it was not until 1968 that the United States Supreme Court struck down a Scopes-era Arkansas law that had been modeled on the Tennessee statute. The Court held in *Epperson* v. *Arkansas*[38] that the "First Amendment does not permit the state to require that teaching and learning must be tailored to the principles or prohibitions of any religious sect or dogma." Arkansas had violated the Establishment Clause that proscribed it from enacting into state law "the religious view of some of its citizens" by prohibiting the teaching of evolution.

Taking a chapter out of the history books, the faithful figured that if they could not be the star they would at least seek to share the stage. In 1982, for instance, Louisiana passed a law, the Balanced Treatment for Creation-Science and Evolution-Science in Public School Instruction Act, which provided that evolution could not be taught in Louisiana public schools unless accompanied by instruction in creation science. Under the statute, schools could decide not to teach both evolution and creation science; but if they taught that man might have evolved from apes then they had to teach that man might have been created by God. Under the act, either both sides got on stage or the show did not go on. The Louisiana law was challenged by parents, teachers, and many religious leaders as an unconstitutional establishment of religion by the state. The case reached the United States Supreme Court in 1987.

In *Edwards* v. *Aguillard*,[39] the Supreme Court, with Justice Brennan writing for the Court, struck down the Louisiana law on the intuitively sound but constitutionally dubious basis that Louisiana's primary purpose was to advance religion. It was constitutionally dubious because there was little real evidence in the legislative record to support this conclusion. In fact, proponents of the law had assiduously avoided making any statements that would make the law appear religiously motivated. The Court's conclusion was based, as Judge Gee of the Fifth Circuit Court of Appeals stated in dissent to a similar holding in the lower court, on "its visceral knowledge regarding what must have motivated the legislators."[40] The avowed purpose of the law, as explicitly stated in the statute itself, was "protecting academic freedom." The legislative history, to the extent any existed, largely supported the conclusion that this was the motivating purpose behind the law.

In assuming that Louisiana had ulterior, sinister, motives in enacting the law, Justice Brennan ignored traditional principles of constitutional interpretation that mandate that a legislature be presumed to have acted constitutionally. It is a basic and long-standing principle that the Court owes due deference to the more political branches of government. And Justice Brennan's opinion ranked high on the constitutional dubiousness scale for a couple of additional reasons. His opinion departed from precedent in finding that a constitutionally defective purpose alone would invalidate the law. Under the prevailing test, the Court had previously always considered whether the law excessively entangled the state in the affairs of religion or had the effect of advancing religion. In other words, in the past, the Court would also have had to find that the law entangled government in religious affairs or advanced religion before concluding that a constitutional violation of establishing religion had occurred. The fact that governmental motives happen to coincide with religious principles is an unsteady altar on which to place a constitutional violation. The fact that a legislative policy is strongly supported by religious groups does not mean it should be suspected of unconstitutionally advancing religion. Such an approach would have doomed abolitionist legislation of the nineteenth century and many humanitarian laws of the twentieth century.

It appears, then, that the Court believed that the Louisiana legislature acted out of religious fervor because it was convinced there was no scientific content to creationism. The logic is straightforward: Louisiana could have had only two purposes in passing the law, secular or sectarian. Since creation science is not real science, then there is no secular purpose. Therefore, Louisiana must have been motivated by a sectarian purpose. The whole argument depended on the Court's showing that creation science was no science. But the Court never made this demonstration.

Justice Scalia wrote a stinging dissent in which he accused the majority of narrow-minded illiberality for rejecting creation science without considering the evidence for it:

> [We cannot] say (or should we say) that the scientific evidence for evolution is so conclusive that no one could be gullible enough to believe that there is any real scientific evidence to the contrary, so that the legislation's stated purpose must be a lie. Yet that illiberal judgment, that Scopes-in-reverse, is ultimately the basis on which the Court's facile rejection of the Louisiana Legislature's purpose must rest.[41]

Scalia is correct that unreflective rejection of one proposed scientific explanation is as bad as unreflective acceptance of another.

In *Edwards*, Louisiana defended its law on the basis of two claims. It first defended creation science as a bona fide science. The basic theory holds that life appeared abruptly in the fossil record and has remained relatively static throughout time. As a legitimate scientific theory, Louisiana asserted, creation science could be taught without appeal to the Bible. According to the "hundreds and hundreds" of scientists cited by the state, creation theory accords better with the facts than does evolution. Moreover, since there were only two viable theories of life's origins, teaching creation science would highlight the flaws allegedly replete in evolutionary theory.

This claim for the validity of creation science led into the second prong of attack, a challenge to the viability of evolution. Louisiana asserted that evolution was not a "fact" and, indeed, might be better described as a "guess" or even a "myth." As myth, it has become a central tenet of "secular humanism," which the Supreme Court has described in other cases as tantamount to "religion." Through this reasoning, the teaching of evolution is itself unconstitutional as a violation of the Establishment Clause.

In *Edwards*, the Court missed an invaluable opportunity to make a statement about its and the Constitution's commitment to science. In fact, a group of Nobel laureates filed an *amicus* ("friend of the Court") brief that sought to demonstrate the scientific foundation for evolution and the lack of any similar grounding for the theory of creation. Despite this expert assistance, the Court seemed fearful of substantively evaluating the scientific merit of creation science. In the opinion, it avoided the subject entirely. I suspect that this was due to the Justices' insecurity in their knowledge of science. In short, the Court was fearful of examining whether creation science or evolution were "sciences" out of concern that they would not be able to recognize science when they saw it. Although the Nobel laureates' brief would have made this task simple enough, it would have established a troublesome precedent. Henceforth, the Court would have been obliged to distinguish between good and bad science, at least in constitutional cases. Instead, the Court relied on its own prejudices regarding creation science and assumed that the Louisiana Legislature must have been motivated by a desire to advance religion.

But Justice Scalia also did not engage in a meaningful examination of the methodological rigor of creation science. He sought to demonstrate merely that Louisiana might have acted in good-faith pursuit of educational goals and was not motivated primarily by religion. He was, for instance, impressed with the testimony before the Louisiana Legislature on behalf of the law. It was, Scalia observed, "devoted to lengthy and, to the layman, seemingly expert scientific expositions on the origin of life."[42] Particularly impressive, he noted, was the fact that this testimony "touched upon" a broad range of subjects, including "biology, paleontology, genetics, astronomy, probability analysis, and biochemistry."[43] But Scalia, like Brennan, never seriously parsed the scientific evidence for the theory of creation and so could not say whether Louisiana's stated purpose for the law was legitimate. Although the Constitution expects the Supreme Court to be deferential to state legislatures, it does not require it to be a dupe for them. As "a layman," Scalia was willing to be seduced by the "quite impressive academic credentials" of the state's scientists. For Scalia to believe that Louisiana acted in good faith means that he found some scientific grounding for creation science. Rather than consider whether this was truly so, Scalia simply abdicated any responsibility for this inquiry to the state's experts.

Edwards illustrates a trend we will confront throughout the sundry topics in the pages ahead. The Supreme Court and courts generally are reluctant to delve too deeply into scientific matters. This insecurity with science, however, has real costs. In particular, it creates an assortment of doctrinal problems for the law, as justices and judges do somersaults to avoid substantive scientific analysis. In *Edwards,* for example, it led the Court to distort its own First Amendment jurisprudence in relying disproportionately on legislative purposes, purposes not expressed in the legislative record. This rendered the decision weaker than it could have been and made it a precedent that might create problems in future cases. The real reason the Court invalidated the Louisiana law was that it did not believe creation science was a science, so the legislature's stated motives were a lie. The Court's ignorance of science caused it to miss an invaluable opportunity to demonstrate this lie by showing that creation science is a fraud.

The Law Hydra

The Law's Science

Values, or policies, are the ultimate currency of the law. The law needs science to help it know about the facts of the world in which legal policy must operate. Without such knowledge, legal policy is literally blinded. Yet, at the same time, policy sometimes dictates certain assumptions about the real world. In other words, the law has its own version of science that occasionally dictates descriptions of how the real world works—whether it works that way or not. In particular, the law's version of science often has very different starting assumptions than science's version of science. There is no clearer example of this than the law's assumption that people have free will and science's assumption that behavior is determined by some combination of nature and nurture. These alternative assumptions serve the two professions well. The law assumes that people are "responsible" for their actions, and, with very few exceptions, it is merely looking for those responsible in order to hold them accountable for their behavior. Science assumes that people are affected by their biology and their experiences, and it is merely looking for the variables that account for their behavior. While these differing starting premises serve the two professions well in their separate capacities, they create intellectual chaos when the two fields must work together. This chaos is especially well illustrated by the context known in the law as "insanity" and in science as "mental illness."

Consider the following ugly facts. In 1984, Leroy Hendricks, a fifty-year-old man, was tried for the molestation of two thirteen-year-old boys.[44] In the actual case, he was convicted of the substantive offense of taking "indecent liberties" with two minors and was sentenced to ten years in prison. Imagine that Hendricks's defense had been insanity. Hendricks testifies that he cannot control his urge to touch the boys sexually. Moreover, when he suffers stress, he is unable to control the urge to engage in sexual activity with children. "I can't control the urge when I get stressed out," he testifies. He understands that his behavior causes grave harm to the children and that "it is wrong." But, when asked how he could be stopped from molesting children, he states, "The only way to guarantee that is to die."

Hendricks thus claims that he is "mentally abnormal" in that he cannot control his behavior. Indeed, in support of this claim he introduces evidence that he has a history of molesting children that dates to 1955, when he was twenty. At that time he exposed himself to two young girls. He has been in and out of prisons and mental hospitals since his first conviction in 1957 for lewdness for playing strip poker with a fourteen-year-old girl. Some of these convictions involved children as young as eight. In 1973, shortly after being released from prison, he began molesting his own stepdaughter and stepson, which lasted five years, though he was never tried for these offenses.

Hendricks calls as an expert a forensic psychiatrist who testifies that Hendricks is a pedophile who cannot control his behavior. In fact, the state's own expert testifies that Hendricks is a pedophile who cannot control his behavior. The state's expert, however, believes that Hendricks does not have a "personality disorder" nor is he "mentally ill" as that term is commonly used. Finally, Hendricks fully understands the consequences of his behavior and the fact that it is both wrong and illegal.

Should Hendricks be "acquitted" on the basis that he is insane? Of course, such an outcome does not mean that he goes free—far from it. A verdict of insanity brings with it an indefinite commitment to a mental institution. In any case, the question in most American jurisdictions, including Kansas, where Hendricks was tried, is moot. In most states, to assert a defense of insanity the defendant must claim that as a consequence of his insanity he could not distinguish right from wrong. Hendricks, therefore, could not even raise a claim of insanity, since he knew that what he was doing was wrong.

This modern right/wrong test is based on the nineteenth-century trial and acquittal of Daniel M'Naghten.[45] M'Naghten had been under the delusion that Sir Robert Peel, the British prime minister, was persecuting him. On January 20, 1843, in an attempt to assassinate Peel, M'Naghten shot and killed Peel's assistant, Edward Drummond, by mistake. However, because he was delusional, M'Naghten was acquitted on the ground of insanity. The British public exploded with outrage, and Queen Victoria demanded that the House of Lords summon the common law judges to explain the result. In response to this summons, the judges framed what has come to be known as the *M'Naghten* test. The judges explained that a defendant should be acquitted if he "was laboring under such a defect of reason, from disease of the mind, as not to know the nature and quality of the act he was doing, or, if he did know it, that he did not know he was doing what was wrong."[46]

The main modern competition to the *M'Naghten* test for insanity is the standard set forth by the prestigious American Law Institute (ALI). This test was intended to be more practical than *M'Naghten*: "A person is not responsible for criminal conduct if at the time of such conduct as a result of mental disease or defect he lacks substantial capacity either to appreciate the criminality [wrongfulness] of his conduct or to conform his conduct to the requirements of law."[47] The ALI test combines the cognitive component of the *M'Naghten* test with a volitional component or irresistible impulse test. Since Hendricks claimed that his behavior was an irresistible impulse, the ALI test would have been available to his defense. But the irresistible impulse component has been severely criticized. In particular, critics complained that it was impossible to distinguish an irresistible impulse from an impulse simply not resisted. In addition, mental health experts objected that it suggested a compartmentalization of the cognitive and volitional parts of the brain. Nonetheless, the ALI test achieved substantial success at first, especially in the federal courts. This success ended abruptly, however, with John

Hinkley's acquittal under the test. The Hinkley acquittal led to congressional passage of the Insanity Defense Reform Act, which essentially abolished the volitional prong and returned the federal courts to the *M'Naghten* test.[48]

According to the law, then, Leroy Hendricks was sufficiently responsible for his behavior to be punished by being sent to prison. After languishing in prison for ten years, Hendricks was about to return to society in 1994. The State of Kansas, however, had other ideas. As his release date neared, the state sought to civilly commit him as a "sexual predator" under the Kansas Sexually Violent Predator Act.[49] Now the roles had reversed. The state sought to prove that Hendricks "suffers from a mental abnormality or personality disorder which makes him likely to engage in predatory acts of sexual violence." This mental defect made him dangerous enough to "engage in predatory acts of sexual violence or sexual activity with children if permitted to do so." The state now took the position that Hendricks had no control over his behavior and had to be committed for the safety of the community. The defense now had to call into question the validity of the psychiatric evidence. At the commitment hearing the defense introduced evidence that indicated that reoffense rates for sex offenders range from 3 to 37.5 percent for those who received treatment and from 10 to 40 percent for those who received no treatment. The defendant's psychiatrist also testified that psychiatric predictions of violence are notoriously unreliable. Hendricks lost at this hearing, and he was incarcerated indefinitely in a secure mental hospital.

Hendricks appealed. His commitment was overturned by the Kansas Supreme Court on the basis that the Kansas Sexual Predator Act failed to require that the person being civilly committed be found to suffer from a "mental illness" and thus violated the Due Process Clause of the Fourteenth Amendment.[50] Relying on a 1992 Supreme Court decision, *Foucha v. Louisiana,*[51] the Kansas Court held that due process requires that a person be both mentally ill and dangerous before he can be involuntarily civilly committed. *Foucha* seemed to have established a level of correspondence between legal "insanity" in criminal cases and "mental illness" in civil commitment determinations. In *Foucha*, the defendant had been found not guilty by reason of insanity. Subsequently, however, he regained his sanity when he recovered from what was "probably" a "drug-induced psychosis." Louisiana sought to keep him incarcerated until he could prove that he was no longer dangerous. The Court concluded that dangerousness alone was not sufficient. Due process allows an insanity acquittee to be incarcerated only "as long as he is both mentally ill and dangerous, but no longer."[52] The Court explained that an insanity verdict "establishes two facts: (i) the defendant committed an act that constitutes a criminal offense, and (ii) he committed the act because of mental illness."[53]

Inherent in *Foucha*, however, is a basic tension between the law's "insanity" and science's "mental illness." It is probably true that all those found to be insane suffer from mental illness, but not everyone suffering from mental illness would qualify as insane. Insanity is a political compromise that offers a rough balance between traditional legal principles of accountability and the need to protect society. Mental illness is a political compromise of a different sort, reflecting both scientific knowledge of human behavior and professional interest in providing treatment for those who can benefit from it (and preferably have insurance to pay for it).

Kansas v. *Hendricks*[54] reached the United States Supreme Court in 1997. The Court held that the Kansas law was not unconstitutional. Justice Thomas, who had dissented in *Foucha*, wrote for the Court, with Justices Breyer, Stevens, Souter, and Ginsburg dissenting from the holding and various parts of the reasoning. At least eight of the Justices agreed, however, that the Kansas law satisfied "substantive due process" requirements under the Fourteenth Amendment. Five members of the Court also found that the law did not violate either the Double Jeopardy or Ex Post Facto Clauses, a conclusion with which the four dissenting Justices disagreed.

Justice Thomas found that *Foucha* had established no categorical due process requirement that commitment be premised on both mental illness and dangerousness. Instead, *Foucha* stood for no more than that proof of dangerousness must be coupled "with the proof of some additional factor."[55] The Court concluded that "mental abnormality" could suffice, since "it narrows the class of persons eligible for confinement to those who are unable to control their dangerousness."[56] Hendricks's mental abnormality meant that he lacked "volitional control."[57] Justice Breyer, agreeing with the majority on this issue, summarized the Court's reasoning:

> Hendricks's abnormality does not consist simply of a long course of antisocial behavior, but rather it includes a specific, serious, and highly unusual inability to control his action. (For example, Hendricks testified that, when he gets "stressed out," he cannot "control the urge" to molest children. . . .) The law traditionally has considered this kind of abnormality akin to insanity for purposes of confinement.[58]

The *Hendricks* Court thus seems to have reinserted the volitional prong into the legal standard. This, of course, sets the Court's commitment jurisprudence at odds with how most jurisdictions approach insanity, including federal courts. Indeed, it might even make commitments under sexual predator laws very difficult, given the extraordinary difficulty in practice of distinguishing between an irresistible impulse and an impulse not resisted. But before we conclude that the Court has shrunk the states' net in commitment cases or widened the net for defendants to claim insanity, we should reflect for a moment. The *Hendricks* majority—Thomas, Rehnquist, Scalia, O'Connor, and Kennedy—are not known for their bleeding hearts. It is exceedingly unlikely that this lineup believed that if Hendricks were to get out and commit another offense his lack of volitional control would excuse his behavior. Nor is it terribly likely that states will have to prove that every person they seek to commit as a "sexual predator" lacks volitional control. What the Court really did was to widen the net considerably to permit states to catch "sexual predators" before they can prey, especially on children. Although *Hendricks* exemplifies a striking discontinuity between the criminal and civil dockets in terms of philosophies of human behavior, the legal result is consistent at a base practical level: on either side of the docket, these people get locked up.

The question, then, is just how wide the new net is. Or just who "these people" who are to be locked up before they have done anything wrong are. If, in fact, lack of "volitional control" is not a prerequisite to commitment in Kansas or elsewhere, then only a finding of "mental abnormality" is required. But what does this term mean? Without the volitional component, the "additional factor" of mental abnormality appears to have no content. It seems to be whatever the legislature says it is. The Kansas statute defines mental abnormality as

> [a] congenital or acquired condition affecting the emotional or volitional capacity which predisposes the person to commit sexually violent offenses in a degree constituting such person a menace to the health and safety of others.[59]

In effect, a person is mentally abnormal because he is dangerous. Upon close analysis, then, under the Kansas statute a person can be committed if he is both dangerous and dangerous. Thus, there is no "additional factor" after all.

"Dangerousness," or a prediction of violence, therefore, is the foundation of the decision as well as the key to satisfying substantive due process. Surprisingly, however, the Court gave it little attention. And the Court said *nothing at all* concerning the uncertainties associated with predictions of violence. It turns out that predictions of violence are untrustworthy. In fact, even when made from among groups with the highest rates of recidivism, predictions of violence are wrong more often than they are correct.[60]

Given that "dangerousness" is the essential factor, might states seek to incarcerate individuals who are deemed dangerous but have never committed any offense? Suppose in the future some set of genetic markers indicates that people with those genes are "predisposed" to "commit sexually violent offenses in a degree constituting such person a menace to the health and safety of others." Could he be locked up constitutionally? In *Hendricks*, the Court noted that the Kansas statute requires that the person "has been convicted of or charged with a sexually violent offense"[61] and thus, in the Court's words, "requires evidence of past sexually violent behavior."[62] Hendricks himself had been convicted of a long series of offenses. But since the statute also applies to those who have only been "charged with" a sexually violent offense, evidence of "past sexually violent behavior" will likely be by proof less than beyond a reasonable doubt. Moreover, prior convictions apparently are not required by due process. In fact, a prior conviction requirement would be theoretically inconsistent with the Court's conclusion that the Kansas statute does not violate the Double Jeopardy and Ex Post Facto Clauses. Hendricks was being incarcerated for what he might do, not for what he did do. Committing only those previously convicted might suggest that the legislature's intent was punitive.[63] Such a punitive intent, the Court said, would be unconstitutional. So long as there is sufficient evidence to find the person is dangerous, the Constitution does not bar civil commitment.[64]

So should the Hendrickses of the world be locked up forever? Should genetic tests be developed someday to identify those predisposed toward sexual violence? Science cannot answer these questions. We might wish that they could be answered in black and white, but both law and science confront a world that comes in many shades of gray. Science tells us that we will be wrong most of the time when we send the Hendrickses of the world to rot in a prison or mental hospital. Genetic tests don't give individuals the opportunity to "beat the odds." The value of liberty is impossible to quantify, but is clearly cherished by our society. The costs of liberty, the costs of permitting the Hendrickses of the world to return to the street (up to 37.5 percent of whom will commit further crimes), are equally difficult to assess. Sexual assaults against children are the hardest crimes to comprehend and the kinds of crime that understandably provoke the most vehement public response. Society might be safer if all those who were "likely to be violent" were locked up forever. But is it a society worth saving?

The Cathedral of the Law

The law is the witness and external deposit of our moral life. Its history is the history of the moral development of the race.

—OLIVER WENDELL HOLMES

The law, for its part, is inherently concerned with both the why and the how of the world; law is, by definition, both descriptive and prescriptive. At bottom, the law is a governing institution that must understand what is; it is also specifically designed to say what ought to be. These tasks require an intimate understanding of nature that is constantly informed by norms and ideals. In fact, the law could be defined simply as humankind's attempt to understand and control nature (including, especially, human nature) to fulfill God's ideal, however God might be defined.

The law has retained a strong religious identity in at least two very distinct respects. The first concerns how law in the United States has assumed near religious significance itself. This stems from its primary societal role in defining the boundaries of acceptable behavior and in meting out punishment for transgressions to the established moral order. An illuminating instance of this religiosity of law is the fairly common observation that the Constitution has achieved scriptural significance: reading its text is akin to biblical interpretation. The second religious aspect of law has somewhat greater historical significance and is particularly related to several themes of this book. The law sometimes incorporates into its operating premises a religious understanding of the universe. This was especially true historically, when the law occasionally assumed some of the more intolerant beliefs of religious practice.

As noted, the law inevitably must rely on factual descriptions of the world in crafting legal prescriptions for appropriate behavior. In a prescientific world, religion offered such descriptions. Where religion leaves off and science begins is a question the law has not always handled with precision. An excellent illustration of the law's difficulty in this task comes from a particularly salient tragedy in the summer of 1692. The witch trials that blazed through Salem, Massachusetts, that summer were solidly grounded in the science of the day.[65] But it was a science based on an empiricism informed through religion's looking glass. The devil and his disciples were real and their influence could be deduced through the effects they caused. At least these effects were known with enough confidence to condemn and execute nineteen people.

The curse first befell Salem the previous winter when Tituba, a Carib Indian slave owned by the Reverend Samuel Parris, entertained

Parris's nine-year-old daughter, Betty, and his niece, eleven-year-old Abigail Williams, with fortune-telling and magic. Soon, Betty and Abigail invited eight other girls to join the fun. The fun turned sinister, however, as Betty and Abigail began to feel guilty about "practicing magic." Shortly thereafter, Betty grew ill. But her illness bore ominous portents, for she exhibited very strange behaviors, falling into trances, screaming blasphemously, and suffering convulsions. Abigail and some of the other girls started to display similar symptoms.

Dr. William Griggs examined the girls and concluded that they were bewitched. Under considerable pressure to name names, Betty and Abigail identified three people: Tituba, Sarah Good, and Sarah Osborn. Betty explained to her parents that she and the other girls had been approached by the devil—a man in black—and offered riches. When they refused his overture, they were set upon by three witches. On February 29, 1692, these alleged witches were arrested.

What began as a small, albeit dangerous, fire received its first dose of gasoline with the confession of Tituba. Although she initially denied any complicity with the devil, under repeated questioning she confessed to practicing witchcraft. She testified to having seen the devil, who appeared to her "sometimes like a hog and sometimes like a great dog." Tituba told the investigating magistrates: "There is four women and one man, they hurt the children, and then they lay all upon me, and they tell me, if I will not hurt the children, they will hurt me." Tituba further told of the spectral evidence of "talking cats, riding on sticks, and a tall, unidentified man of Boston." Finally, and most ominously, she told the magistrates that there was a conspiracy of witches in Salem.

In the ensuing weeks, many people came forward to complain that they too had been victimized by witches or had seen spectral shapes of some of their neighbors. Joining the principal accusers of Betty Parris and Abigail Williams were Ann Putnam and Elizabeth Hubbard. Many of the first to be denounced were women whose behavior or economic circumstances somehow disturbed the social convention. An early writer's description of the typical witch sounds just out of central casting: "a hagged old Woman, living in a little rotten Cottage, under a Hill, by a Wood-side, and must be frequently spinning at the Door: she must have a black Cat, two or three Broom-sticks, an Imp or two, and two or three diabolical Teats to suckle her Imp."[66] As many contemporary commentators have pointed out, the witch trials were as much a pogrom against women as a holy war against the devil.[67] In particular, women who reveled in "infirmity, impotence of passions and affection, . . . and vagrant lust" were "the fittest subjects" for the devil.[68] Richard Baxter, an influential Puritan scholar, described the gendered nature of witchcraft as follows: "Lustful, rank girls and young widows that plot for some amorous, procacious design, or have imaginations conquered by lust . . . [there] Satan oft sets in."[69]

The list of the accused, however, soon exceeded the likely suspects from the rolls of the destitute and ignorant and encompassed both men and women of some means and community standing. Four-year-old Dorcas Good was jailed after several witnesses testified that the child's specter fell upon them and bit and pinched them. Dorcas eventually confessed, and she was spared the fate of her family. Her baby sister died at her mother's side and her mother joined the ranks of the martyrs at the gallows. Dorcas reportedly went insane.[70]

The growing list focused on Quakers and other "social deviants" who populated the area, including such eminently respectable citizens as George Burroughs and Rebecca Nurse. In addition, the devil seemed to have a particular proclivity to enlist as witches opponents of the Putnam family of Salem Village, an old and well-established family that resented many of the changes occurring in the village and town. By the spring, the band of accusers had even identified John Alden, a well-known sea captain and merchant, as a witch, though he had to be pointed out to the girls. Abigail Williams went so far as to accuse Samuel Willard—a prominent Boston minister and critic of the trials—until she was whisked out of court and firmly informed that she was mistaken.[71] She got the hint and withdrew the accusation.[72]

But a question that redounds through the centuries is what proof there was that the accused were witches. The evidence lay in a mix of common knowledge and empirical fact. For instance, it was commonly known that witches targeted children.[73] But, as Michael Dalton's much relied-on work, *The Country Justice* (1619), pointed out, since the witch's world was the world "of darkness," no "direct evidence" would prove their presence. Instead, indirect and circumstantial proof had to be relied on. There was plenty of this kind of evidence to be found, if only the magistrates looked carefully enough. For instance, witches typically had unusual marks on their bodies, so-called witches' teats. According to theory, the witch's "familiars" needed a place to suckle. In practice, virtually any strange mark would suffice. In little Dorcas Good's case, a red mark "about the bigness of a flea bite"—which it probably was—was sufficient. Of course, absence of marks would not exonerate an accused, as Rebecca Nurse discovered to her chagrin. She suffered through repeated physical examinations in search of telltale marks, all of which ended inconclusively. She was convicted and hanged nonetheless. Another preferred test for detecting witchery was the "touching test." The judges sometimes required the defendants to touch the afflicted witnesses and if, in so doing, they alleviated the suffering, they thereby proved that they were the cause of it.

Showing their enlightenment, however, the judges did eschew certain barbaric forms of proof. In particular, they rejected such classic tests as placing the suspected witch in water to see if she would float or burning her. Such "ordeals" were deemed uncivilized, and the judges sought more scientific alternatives.

The seventeenth century was a time when thinking was becoming more scientific, more rational, and some of the most respected thinkers on demonology and witchcraft were otherwise modernists in their science. Principal among these commentators was Cotton Mather, son of Increase Mather, and a careful researcher of specters and other devilish goings-on. Mather very much considered himself a scientist and, indeed, was a member of the prestigious Royal Society. Mather had studied victims said to be bewitched firsthand and had no doubt that the phenomenon was "dreadfully real."[74] The logic of causation he employed was also rigorous. If a person quarreled with a witch and then became ill or some misfortune befell the person, it could be deduced that the witch bore responsibility for the misfortune. As a legal matter, if the witch used spectral visitations to accomplish her work, then she could be accountable for battery. If death resulted, she was guilty of murder. In any event, in Massachusetts the law required no direct harm to be done: "If any man or woman be a witch (that is hath or consulted with a familiar spirit) they shall be put to death."[75] This edict was based squarely on biblical teachings: "Thou shalt not suffer a witch to live."

In the case of Bridget Bishop, for example, the evidence was overwhelming. Befitting the strong proof available against her, Bishop was the first to be tried. Confessed witches Deliverance Hobbs and Mary Warren both testified that Bishop was one of them.[76] A group of matrons inspected Bishop's body and found an "excrescence of flesh . . . not usual in women."[77] Most damning, however, was the scientific proof of "spectral evidence." Her accusers testified that it was Bishop's specter that attacked them. A surfeit of other witnesses also spoke of how Bishop's specter had appeared to take credit for the many misfortunes they had suffered. On one occasion, a witness related, Bishop "was under a Guard, passing by the great and spacious Meeting-house of Salem, she gave a look towards the house, and immediately a daemon invisibly entering the Meeting-house, torn down a part of it."[78] This was enough for the jury. Bishop was hanged on June 10.

Spectral evidence was, by far, the most controversial form of scientific proof introduced in the Salem trials. Specters of the accused would be seen doing evil deeds or consorting with the devil. This form of evidence, of course, vitiated any alibis that might be forthcoming. But highly disputed among the scientists and ministers was whether the devil could use the shape of a good person in this way. If not, then spectral proof might be enough to support a conviction. Yet many, including Cotton Mather, urged that there was no proof that the devil could not take a good person's visage for evil, and the Bible contained examples of good men possessed by evil spirits. Mather, therefore, cautioned the Salem judges to require other evidence before convicting an accused.[79] This warning, however, was not always heeded in the hysteria of the time. It was Cotton Mather's father, Increase Mather, who wrote the most damning critique of spectral evidence, in *Cases of Conscience*, which led to the abandonment of this form of proof. Although Increase thought witches probably existed, he considered spectral evidence too prone to abuse, since the devil himself might employ specters to ensnare the innocent. Demonstrating a modicum of the rationality that would dominate the next century, Increase wrote in 1692:

> "Presumptions . . . whereupon persons may be Condemned as Guilty of Witchcrafts, ought Certainly to be more considerable, than barely the Accused Persons being Represented by a Spectre unto the afflicted; inasmuch as 'tis an undoubted and a Notorious Thing, That a Daemon may, by God's Permission, appear even to Ill purposes, in the Shape of an Innocent, yea, and a vertuous man."[80]

Logic dictated, after all, that the judges should beware of being "frequently Liable to be abused by the Devils Legerdemains."[81]

Modern courts have not been entirely guiltless in conducting their own witch-hunts. Although not specifically based in religion, as were the witch trials of 1692, the two most often cited contemporary versions have involved repressed memories and child sex abuse trials. The former relies on the problematical theory that memories can lie repressed for decades and be "recovered" intact through psychotherapy or hypnosis. And many child sex abuse cases were built on the testimony of young and impressionable witnesses who were subjected to leading questions and significant pressure to "tell the truth." In these cases, the "helping professions" assumed the priestly task of rooting out evil incarnate. In most cases, the rational thinking displayed by Increase Mather would go a long way toward avoiding many of the witch-hunts the law occasionally plunges into. The value of the scientific method is not always in the findings it provides but in giving us a critical perspective regarding findings that are claimed but just do not add up.

Science, however, is no panacea. Indeed, most witch-hunts are carried out in the name of science. But it is typically science that is ill-conceived and wielded with great passion and prejudice. A more sober and sophisticated understanding of science should at least temper these flames somewhat.

Law, science, and religion will always lie in uneasy tension with one another, for they share too many objectives and too few working premises. This tension, however, is not altogether bad, for it operates to check their power. Nothing would be more frightening than to have one of these institutions too fully dominate the other two. For the law, the objective must be to work with science and religion but never to be dominated by them. This can only come through knowing them well. As Vito Corleone said in Francis Ford Coppola's *The Godfather*, "Keep your friends close, but your enemies closer."

II

An Overview

How Law and Science Meet — From Courts to Congress

Laws and institutions must go hand in hand with the progress of the human mind.

— Thomas Jefferson

Applied science is a conjurer, whose bottomless hat yields impartially the softest of Angora rabbits and the most petrifying of Medusas.

— Aldous Huxley

T heirs was a marriage that divorce courts see almost every day. When they first met, Mary Sue Davis was nineteen and Junior Lewis Davis was twenty-one. Their marriage lasted nine years. Ironically, as it turns out for our story, he worked as a refrigeration technician; she was employed as a sales representative. Together they earned approximately $35,000 a year in 1989, the year of their divorce. They had accumulated a modest amount of property, but not enough to fight over. Ordinarily, their divorce would have been a routine matter, at least as far as such things can ever be described as routine. There was, however, one issue that the Davises could not settle. Sitting in a freezer in the Fort Sanders Medical Center in Knoxville, Tennessee, were seven cryopreserved embryos that belonged to them both. The dispute over this biological potential would catapult the Davises to the forefront of the burgeoning field of biotechnology regulation. *Davis* v. *Davis*[1] was the first case of its kind and remains the benchmark by which similar cases are evaluated. The case, or at least the trial court portion of it, also stands as a symbol of how the complexities of science can make difficult legal problems even more difficult. The lessons of *Davis* v. *Davis* are manifold: it reveals the assorted tensions between advancing technology and settled principles of law; it frames the interconnectedness of the various lawmaking bodies in the United States—courts, legislatures, and administrative agencies; it illustrates the corrupting influence that high-profile, media-hyped stories can have on the decision process; and it portrays the basic human component that lies at the base of virtually all the tales where law and science meet.

The Davises desperately wanted to have a child. Mary Sue suffered five tubal pregnancies, the first of which resulted in surgery to have her right fallopian tube removed. After the fifth tubal pregnancy, she had her left fallopian tube ligated, which left her without functional fallopian tubes by which to conceive naturally. Desperate and not wealthy, the Davises decided to try in vitro (literally, "in glass") fertilization through Dr. Ray King's Knoxville clinic. After six months of frustrating failure with the in vitro process, the Davises pursued still another option. Their fortunes seemed to brighten when they were successful in arranging to adopt a child. But at the last moment, the child's birth mother changed her mind and refused to put the child up for adoption. Clearly, the fertility Gods had failed to smile on the Davises. Mary Sue and Junior returned to Dr. King's clinic.

By this time, in 1988, Dr. King's clinic employed the new cryopreservation technique, which constituted a marked improvement over past practice. In their initial attempts in 1985, the Davises suffered through the in vitro procedures six times, at a total cost of $35,000, but without success. Each attempt involved a month of injections necessary to shut down her pituitary gland and eight days of injections necessary to stimulate her ovaries to produce ova. Mary Sue later testified to her deep fear of needles. Further, she had to be anesthetized for the aspiration procedure to be performed, and forty-eight to seventy-two hours later she had to return to the clinic to have the embryo implanted in her uterus. Junior's task, donating his sperm, was rather less unpleasant, and he faithfully stayed by Mary Sue through "the many anxious hours" while she underwent the aspiration and implant procedures. After each procedure, Mary Sue and Junior would await news that never turned out to be good. Cryopreservation, however, eased the hardship considerably. It permitted extraction of several ova at once, which then could be inseminated together in the laboratory and allowed to develop as fertilized zygotes to the point at which they would be mature enough to be implanted into Mary Sue or cryopreserved for future implantation. The pain and anguish of the aspiration procedure did not change with cryopreservation, but they did not need to be endured quite so often.

The last time Mary Sue underwent aspiration, on December 8, 1988, the doctor retrieved nine ova for fertilization. The fertilized eggs, or zygotes, were allowed to develop in petri dishes until they matured to the four- to eight-cell stage. Two of these were implanted in Mary Sue and the remaining seven were frozen in nitrogen and stored at minus 196 degrees centigrade. Once again, the implantation was unsuccessful. Before the procedure could be repeated, Junior filed for divorce. Throughout the long ordeal, the couple never disclosed the reasons for their separation. The court papers stated prosaically only that they had "irreconcilable differences."

The Davises had not signed any agreement or even discussed what to do in the event of death, divorce, or other similar calamity. Tennessee had no law that applied to this situation, nor had Congress or any federal agency passed guidelines or regulations that might resolve the dispute. In 1988, only one state had addressed this subject. Under a 1986 Louisiana statute, disputes between parties like the Davises are to be resolved in the "best interests" of the embryo.[2] The Louisiana law further mandated that unwanted frozen embryos must be made available for "adoptive implantation." Other than this probably unconstitutional law, the legislative policy arena was, if you will pardon the expression, barren ground. Tocqueville wrote that "scarcely any political question arises in the United States which is not resolved, sooner or later, into a judicial question."[3] In modern times, at least, and especially when the issue is as controversial as cryopreserved embryos, these political questions very often become judicial questions sooner rather than later.

The same legal question usually arises in a variety of legal contexts. When this question has a large science component, the context can matter greatly to how it is answered. In the example of in vitro fertilization, legal decision makers who might address the issue include legislators, executive officers or administrators, and courts. Legislators, state or federal, might establish guidelines that restrict or even prohibit the practice. Federal or state agencies might regulate clinic practices, if within the agencies' jurisdiction, or they might adopt funding guidelines for research on embryos. The courts also might become involved, either to interpret and apply laws and regulations promulgated by legislatures or agencies or to interpret and enforce private agreements between the parties. Finally, some of these issues might rise to constitutional dimensions if, say, claims are made that the woman has a right to procreate, the father has a right not to procreate, or the embryo has a right to life. Although today frozen embryos have received considerable attention at every level of lawmaking, in 1989 the Davises' frozen embryos became a matter for a local trial court to resolve in a divorce settlement.

The Davises and their seven frozen offspring all ended up in Judge W. Dale Young's courtroom in tiny Maryville, Tennessee. The trial received the most press attention of any case in Tennessee since John Scopes had dared to teach evolution in Dayton in 1925. And in terms of legal acumen, scientific sophistication, and media-sycophancy, Judge Young rivaled Judge Raulston of the Scopes trial.

From a legal point of view, the important question, really the only question, concerned the disposition of the embryos between Mary Sue and Junior. The public and media interest focused on a somewhat more dramatic and controversial issue: the nature of the frozen organisms. Judge Young made the mistake of focusing on the latter question. Indeed, he sought to answer the most basic issue of all: "When does life begin?"[4] This was a question, the judge later said, that had been "ducked" by the Supreme Court in *Roe* v. *Wade*[5] He would decide it. His decision, as one commentator later observed, was "a rare combination of hubris and ignorance."[6] Unfortunately, this combination is not that rare, though Judge Young may have displayed an unusual degree of it.

In pursuing the answer to his question, Judge Young set up a false dichotomy. He asked whether the controverted biological material was property or human. Since he found it obviously not property, that left only one possibility: "The Court finds and concludes that the seven cryopreserved embryos are human."[7] This meant that the choice between Mary Sue and Junior hinged on what was in "the best interests" of the "children." Their best interest, not surprisingly, was "implantation to assure their opportunity for live birth." Yet the court never engaged in any substantial analysis of which litigant would better ensure birth. Although Judge Young vested the embryos "in Mrs. Davis for the purpose of implantation," given her prior history it was far from certain that this decision ensured their survival. Moreover, Junior had not specifically advocated destroying the embryos—he just didn't want to be "raped of my reproductive rights"—though at trial he did not have any idea what should be done with them. He opposed Mary Sue's use of them and objected to donating them to anonymous recipients. At trial, he argued that he just wanted the decision to be a joint one. Three years later, when the Tennessee Supreme Court ended this litigation and awarded him custody of the embryos, he destroyed them almost immediately. It is clear that Judge Young understood from the start what Junior's plans were.

Judge Young concentrated all his energies on the scientific questions presented, somehow believing that their answers would dictate the proper legal result. But the issue in the case turned on a debate that has no scientific significance whatever. Much of the expert testimony concentrated on what to call the seven four- to eight-cell organisms, "preembryos" or "embryos." The court proceeded as if medical genetics could inform it whether the frozen entities were or were not yet "human." While such classifications are important for the law, since legal rights might depend on them, such scientific descriptions have no prescriptive force. For example, the law distinguishes "children" from "adults" and, no doubt, science could provide cogent descriptive differences that would help lawyers feel secure in maintaining the categories. But whether these categories should be treated alike or differently is a policy choice that is based on values that in turn make the descriptions relevant. The factual description does not contain an inherent policy recommendation. John Wilke, president of the National Right to Life Committee, illustrated this misconception, shared by Judge Young, when he called the ruling a "progressive decision rooted firmly in the scientific fact that human life begins at conception." Labeling these four- to eight-cell entities "preembryos" or "embryos" was quite beside the point. The point was, given their level of maturity, whether they had any independent rights or, even more to the point, what rights existed for those who claimed them. This is a legal judgment, not a scientific question.

Judge Young, however, thought it was a scientific question and thus brought no legal judgment to bear on the issue. He relied almost exclusively on the expert testimony of Dr. Jerome LeJeune, the director of the French National Center of Scientific Research, the discoverer of the genetic cause of Down's syndrome and an ardent "pro-life" advocate. Dr. LeJeune testified that there was no distinction between preembryos and embryos. Moreover, there was no distinction to be made, in his view, between embryos and ourselves. These seven four- to eight-cell entities were "early human beings," "tiny persons" deserving of full protection; they are our "kin," he testified. He likened cryopreservation to "putting tiny human beings in a very cold space, deprived of liberty, deprived even of time; they are as it were in a concentration camp. . . . It is not a hospitable place as the secret temple of a woman's womb."[8] Finally, he testified—though on what medical basis is not known—that Junior had failed his moral obligation to bring these "tiny human beings to term." In contrast, "Madame, the mother, wants to rescue babies from this concentration can [sic]."[9] Later, Charles Clifford, Junior's attorney, asked LeJeune on cross-examination, "Do you know what this is?" as he held up an egg. "An egg," LeJeune responded. Clifford retorted, "Good. I thought you were going to tell me it was an early chicken."[10]

Just as Dr. LeJeune played the bumpkin in the Davis trial, as William Jennings Bryan had in the Scopes trial, Maryville played host to the circus atmosphere much as Dayton had sixty-four years before. Admittedly, the Davis trial did not rise to the spectacle that enveloped Dayton. It was missing some of the more remarkable personalities, such as Clarence Darrow and H. L. Mencken, the former of whom gave character to the proceedings, the latter of whom brought him to a national audience. Nonetheless, Maryville was aflutter with activity during the three-day trial. Scores of reporters descended on the town. And the town—and the judge—welcomed the attention. T-shirts were hawked on street corners, and the judge even ordered a recess so that the press could attend a brunch in their honor sponsored by the Chamber of Commerce. The judge seemed to fully enjoy his "fifteen minutes" of fame. As he drawled in one interview, "The issues in this case weigh very heavily on my shoulders. I am well aware that the world is watching me now."[11] Unlike some judges who subsequently found themselves in the public spotlight, he did have the good sense to ban television cameras from the courtroom. But he was not media shy. He allowed the television cameras to capture the scene as he handed his written opinion to his clerk.[12] After the decision, he gave interviews widely and often.

The underlying basis for Judge Young's ruling that "human life begins at the moment of conception" was summarily repudiated by both the intermediate court of appeals and the Tennessee Supreme Court. Even Mary Sue's lawyers abandoned this basis in their appeal to the state supreme court, despite their success with it at trial. The Tennessee Supreme Court stated categorically that a "preembryo" is not a "person" under Tennessee law. In fact, it completely rejected the dichotomy established by Judge Young between property and personhood, finding that there was a third possibility. Following the ethical guidelines established by the American Fertility Society, the court held that preembryos "are not, strictly speaking, either 'persons' or 'property,' but occupy an interim category that entitles them to *special respect* because of their potential for life."[13] After that, the Tennessee Court took a classically conservative approach to resolving the dispute. The court asked who, between Mary Sue and Junior, had the "greater interest in the preembryos." Addressing this question had become substantially easier with time.

By the time the case reached the Tennessee Supreme Court, both Mary Sue and Junior had remarried. Mary Sue, moreover, no longer wished to implant the embryos herself. Instead, she asked the court for custody so that she could donate them to a third person. Junior's interests had not changed. He continued to seek custody in order to avoid unwanted fatherhood. Between Mary Sue's interest in donating them and Junior's interest in avoiding fatherhood, the court had little difficulty finding that Junior's interest was greater, and it awarded him custody.

Mary Sue appealed to the United States Supreme Court, but it refused to hear the case. As noted, Junior had the embryos destroyed at his first opportunity.

Testing the Political Winds

Reproductive technologies such as cryopreservation raise profound scientific and moral questions. As a social and political community, we must wrestle with the facts and our feelings in an effort to respond to the challenges raised by these new technologies. This is an issue that transcends scientific contexts and includes such diverse issues as cloning, car emission standards, the survival of the snail darter or spotted owl, genetic manipulation of tomatoes and cows, global warming, and so on. What is especially compelling about these cases is the moral and ethical principles that must be divined in deciding whether to permit cloning or protect the snail darter. They also present substantial intellectual challenges in requiring policy makers to weigh complex scientific information in light of whatever moral or ethical principles are embraced. Yet from a legal (or perhaps political theory) perspective, the compelling issue is not so much what rules are chosen but who chooses the rules. Implicit in the debates about the moral status of new technologies lies an age-old debate from the domain of political theory concerning the nature and legitimacy of government. In the United States, this debate is like a single river with two currents running through it. The currents are sometimes separate, having little effect on one another, sometimes they conflict and might even cancel one another out, and sometimes they combine to constitute a raging torrent. The first current is driven by the division of power in the national government between the legislative, executive, and judiciary branches, what is referred to as the doctrine of separation of powers. The second is the division of power between the national government and the states, known as federalism.

A good illustration of these two currents comes from a legal and scientific question that arises at the opposite end of the spectrum from that posed in *Davis v. Davis*. In the last several years, the American public has been engaged in a substantial debate about "the right to die," including what is generally described as physician-assisted suicide. Interestingly, for lawyers and judges, most of this debate has not been substantive. In other words, the argument is not over the scientific or moral question of determining whether the grim reaper has arrived but instead concerns the procedural question of who decides that the grim reaper has arrived.

Once again, somewhat contrary to Tocqueville's observation that most political questions "sooner or later" become judicial questions, the matter of when death should arrive, like the question of when life has begun, mainly started in court. Back in 1990, the Supreme Court ruled in *Cruzan v. Director, Missouri Department of Health*[14] that states could establish procedural guarantees to ensure that a person who decided to refuse medical treatment had voluntarily and knowingly done so. Nancy Cruzan had sustained severe injuries in a car accident that left her in a persistent vegetative state. Acting in what they deemed were Nancy's best interests, her parents sought to terminate all life support, including all food and water. But under Missouri law, Cruzan's parents could remove life support only if they could demonstrate by clear and convincing evidence that Nancy herself had expressed a desire for this result before the accident rendered her incompetent to make such decisions. Nancy's parents argued that the "clear and convincing" standard erected too high a barrier to the exercise of the fundamental right to refuse medical treatment. The United States Supreme Court disagreed. It conceded that individuals have a right to refuse medical treatment but concluded that a state could require this preference to be clearly articulated by the person asserting it. When the case was later remanded, however, the lower court found that the evidence did clearly demonstrate Nancy's desire to refuse treatment under such circumstances. Her parents authorized removal of the artificial support and Nancy died soon thereafter.

Cruzan posed a difficult case for the Court because of Nancy's inability to express an opinion about her wishes. It was also difficult because the case potentially served as a threshold decision for a host of cases involving individual rights in the context of health, medicine, and death the Court explicitly moved back from the edge and noted that "in deciding 'a question of such magnitude and importance . . . it is the [better] part of wisdom not to attempt, by any general statement, to cover every possible phase of the subject.'"[15] Nonetheless, the *Cruzan* Court recognized for the first time that there exists "a general liberty interest in refusing medical treatment."[16] The definition of the contours of this right was left for another day.

That day seemed to arrive in 1997 when two cases that reached the Supreme Court squarely raised the issue of whether individuals have a constitutional right to physician-assisted suicide. The two cases were *Washington v. Glucksberg*[17] and *Vacco v. Quill.*[18] The basic claims were that state laws that made suicide or assisted suicide a crime violated an individual's rights of liberty and equal protection of the laws. In both cases, the claim was made that a terminally ill, mentally competent adult should be able to choose the moment and manner of his or her death.

The Court concentrated its focus on what it considered to be the state's legitimate interests in determining this moral question. It asserted that states, as they had done here, could legitimately choose to define lives worth protecting in establishing antisuicide laws. The state's interests also included protecting vulnerable groups, such as the depressed or mentally ill, and "protecting the integrity and ethics of the medical profession." At the same time, the Court noted, states were generally free to experiment with alternative schemes that might permit individuals greater control over their final destinies. The Court summarized its position as follows:

> Throughout the Nation, Americans are engaged in an earnest and profound debate about the morality, legality, and practicality of physician-assisted suicide. Our holding permits this debate to continue, as it should in a democratic society.[19]

In its phrase "democratic society," the Court undoubtedly intended to include both currents of American political thought, separation of powers and federalism. What the Court really meant was that this particular political question was not yet a judicial question. But deeming it a political question did not specify which "political" department should decide it. Moreover, that it is a "political question" left to the political rather than judicial departments hardly guarantees its being answered "democratically."

The first current in the stream of American democracy, separation of powers, is the dynamic created by the organization of the federal government. If asked, most Americans would agree with the Supreme Court in describing the two nonjudicial branches, the Congress and the president, as "democratic" institutions. And there is some truth to this perception. But it is also largely inaccurate. To some extent, it depends on how loosely we define democracy. But even at its loosest, it only arguably encompasses the two houses of Congress. Moreover, the Senate is a stretch given that Rhode Island and California each have two senators despite the fact that Californians outnumber Rhode Islanders 32 to 1. So much for "one person, one vote." The executive branch is also only superficially democratic. Although we vote for the president, the outcome is funneled through the archaic mechanics of the electoral college that could permit, and has permitted, someone to be elected president who has received a minority of popular votes. Moreover, the vast and very powerful federal bureaucracy is not elected, and most people have little idea who they are, how they got there, or what they do exactly. The third branch, the judiciary, is generally conceded to be nondemocratic or, even, in Alexander Bickel's words, "countermajoritarian."[20] Judges are nominated by the president and confirmed by the

Senate and serve for life. This process was designed specifically to create a judiciary that was insulated from the whims and the will of the people. Thus, in our entire sprawling federal government, only the House of Representatives can lay claim to being a truly democratic institution. Depending on your point of view, this might or might not affect your continuing support of democracy.

The other main political current running through the debate of who chooses when death has arrived is federalism. Specifically, given that government will play some role in defining the contours of physician-assisted suicide, the next question is which government—federal or state? From the beginning of our republic, the issue of "states' rights" has always been close to the surface. With the Civil War, of course, it boiled over. Most states mirror the federal government's separation of powers, with some state versions being more or less democratic than their federal counterparts. On average, state governments tend to rely on democratic processes much more than does the federal government; many states even elect judges. In the context of physician-assisted suicide, the federalism issue has been a strong force, as is illustrated by the dynamic following the passage and implementation of Oregon's popularly enacted Death with Dignity Act.[21] However, as will become clear as the story develops, the other current, the separation of powers, also runs strongly through this immensely complex and controversial issue.

In 1994, in a popular referendum, the Oregon voters approved the Death with Dignity Act by 51 percent to 49 percent. In fact, they voted twice on the subject, the second time overwhelmingly rejecting an initiative that would have repealed the law. The act gives a competent adult the right to request a lethal dose of prescription drugs if he is judged by two doctors to have less than six months to live. The act remained tied up in court for three years. In November 1997, however, the law went into effect. Almost immediately, Senator Orrin G. Hatch (Utah) and Representative Henry J. Hyde (Illinois), chairmen of the Senate's and the House's judiciary committees, respectively, pressured Thomas A. Constantine, the head of the Drug Enforcement Administration (DEA) of the Justice Department, to nullify the Oregon law. Constantine announced—apparently without consulting his bosses, the attorney general or the president—that the federal government would impose severe sanctions on doctors who prescribed lethal dosages of prescription medication. According to Constantine, the federal drug law preempted Oregon-voters' decision to permit physician-assisted suicide. Although doctors are licensed by the individual states, under federal law they need a special license from the DEA to distribute controlled drugs for "legitimate medical purposes."[22] As Constantine explained, "Delivering, dispensing or prescribing a controlled substance with the intent of assisting a suicide would not be under any current definition a legitimate medical purpose."[23]

On the issue of federalism, the irony here is considerable, since Senator Hatch and Representative Hyde usually associate themselves with strong states' rights positions. It seems that they found a higher principle than states' rights operating here. Senator Hatch defended his apparent hypocrisy by explaining, "Normally, I don't like to interfere with the states. When you're talking about killing people in violation of food and drug laws—and federal laws—then I think that's where you draw the line."[24] Other representatives disagreed, believing this to be the sort of issue that should be the subject of local control. Senator Wyden (Oregon), for instance, made this point bluntly when he complained abut the DEA's interfering in his home state's affairs[25] Senator Wyden explained that he was personally opposed to assisted suicide but argued on the Senate floor that it would now "be wrong for those at the federal level to meddle with that decision." "The people of Oregon voted 'no' on the repeal" of the Act, he said, "and what I'm telling the Clinton administration and the congressional leadership is 'What part of 'no' do you folks not understand?'" He concluded, "It seems to me [that] . . . the Drug Enforcement Administration has plenty to do right now other than to meddle in the internal affairs of the State of Oregon."[26]

After she had had the opportunity to review the federal drug laws, Attorney General Janet Reno agreed in substance with Senator Wyden. In June 1998, she overturned Constantine's statement of DEA policy. She concluded that Congress had passed the drug laws to counter drug trafficking. It did not mean these laws to inject the federal government into the moral and ethical conundrums posed by the Oregon initiative. She pointed out further that earlier in the year Congress had passed and the president had signed the Assisted Funding Restriction Act of 1997, which prohibited the use of federal tax dollars to pay for or promote assisted suicide. That law did not, as one of its sponsors made clear when it was passed, "in any way forbid a state to legalize assisted suicide or even to provide its own funds for assisted suicide."[27] The 1997 law simply denied federal monies to assisted suicide, without prohibiting the practice generally.

Within hours of Attorney General Reno's announcement, a bipartisan bill was introduced in Congress that aimed to reimpose federal control over this divisive issue. The act was entitled the Lethal Drug Abuse Prevention Act of 1998. The basic purpose of the bill was to "clarify" federal law to show that Congress did indeed mean the drug laws to inject the federal government into the moral and ethical conundrums posed by the Oregon initiative. As for federal interference in the states' internal affairs, Senator Nickles offered the following defense:

> I have long been a strong advocate of states' rights and the limited role of the federal government, so let me make clear what this legislation does. It simply clarifies that the dispensing of controlled substances for the purpose of assisted suicide is prohibited under long-standing federal law, the Controlled Substance Act.[28]

In effect, for Senator Nickles, pervasive and long-standing federal interference in a particular policy arena justifies further interference in related policy contexts. I doubt that most states' rights advocates would be reassured by his reasoning.

Like oil and vinegar in salad dressing, law and science seem most settled when left to their separate domains. With some vigorous shaking, however, they will combine, and if in appropriate measures, this mix might even be palatable. But still law and science bear certain fundamental dissimilarities that will always serve to keep them apart. It is worth spending some time here outlining what makes the mix of law and science so difficult. Later, we will see what results when they are mixed, shaken, and poured on the salad we call the United States.

Systemic Challenges

There are essentially four basic contexts in which scientific research enters legal/political decision making: (1) trial and appellate courts in nonconstitutional cases, (2) constitutional cases and especially Supreme Court decisions, (3) legislatures, and (4) administrative agencies. These four contexts pretty much describe the full range of legal users or consumers of scientific research. It is worth pointing out that these four areas are suffused with layers of complexity, each having its own subtle dynamics on the subject of science and law that I can only hint at here. For instance, judges are aided by judicial clerks and sometimes special masters, legislators have bevies of assistants and access to independent governmental bodies (such as the National Academy of Sciences), and administrators have advisory committees and staff assistants. Each legal arena deserves a book of its own and a few have received them.[29] Nonetheless, even when help is available, the principals in these legal contexts—trial and appellate judges, legislators and executive officers—are still responsible for processing scientific information on their way to making policy decisions.

Unfortunately, scientific research does not come in neat prepackaged bundles that can be bought in individually sealed packages at your local supermarket. For the law, a major preoccupation involves determining what constitutes "science" in the first place. Are engineers scientists? What about an odontologist who testifies that the bite mark on the victim's arm matches (or does not match) the defendant's teeth? Historians are probably not scientists, but social anthropologists might be. And what about the physicists and cosmologists who speculate about wormholes or superstring theory? They might be scientists, but if they testified before Congress regarding the value of the superconducting supercollider to their work, could their testimony fairly be labeled "science"?

Even when we can all agree that what we are looking at is "science," when it comes to applying it to legal policy the task becomes exceedingly complex very quickly. Science is essentially a method for producing information, and the sciences use a wide variety of methods in a multitude of contexts. Consider just the single legal matter of the significance of posttraumatic stress disorder to ameliorate guilt in criminal cases. We have an assortment of neuroscience studies that might be done, ranging from animal experiments to postmortem examinations of humans. We might also have psychological studies ranging from clinical case studies to elaborate population studies of naturally occurring events using carefully selected control groups. And among all this research, studies might point in various directions with various degrees of certainty. The difficulty of bringing science to the law should not be underestimated.

There are four readily identifiable and quite substantial barriers to the use of science in every legal context: (1) the availability of data, (2) the lay person's understanding of the science, (3) integrating science into other information, and (4) cultural conflicts between law and science. These barriers are inherent in the two disciplines' different natures and the difficulty presented by their having to work together. The four barriers exist across the legal and political landscapes, though they play out very differently in the various contexts.

Is It Soup Yet? The Problem of Unavailable Data

Perhaps the most basic difference between law and science involves their very dissimilar schedules for decision making. The different timetables of law and science is an issue that has arisen with increasing frequency in recent years. In short, complications arise when the law races ahead of science in deciding empirical questions. These scheduling problems have been most noticeable in tort litigation and federal regulation of allegedly toxic substances such as asbestos, agent orange, bendectin, lead, gulf war syndrome, silicone implants, thalidomide, and so forth. But it also arises in many other regulatory and litigation-related areas, such as those involving the environment, forensic evidence, and the myriad of syndromes and diagnoses generated by psychologists and psychiatrists. The following concrete example should help illustrate the problem.

Suppose you are an attorney, and through your door walks Joseph Client, a thirty-seven-year-old man diagnosed with small-cell lung cancer. He smoked for eight years and both his parents smoked around him throughout his childhood years. For the last twenty years, Client has worked for the city's water and electricity department. This brought him in contact with electrical transformers 40 to 50 percent of the time. Most of these transformers contain a petroleum-based, flammable mineral oil. However, because of the fire danger in some places, some of the transformers were filled with a synthetic fire-resistant chemical that contains polychlorinated biphenyls (PCBs). While testing indicates that over 50 percent of the transformers Client might have serviced are contaminated with PCBs, only 19.2 percent contain levels of PCBs above fifty ppm. The EPA lists a transformer below fifty ppm as a "non-PCB transformer."

Client's lung cancer is quite rare among people his age, even for smokers. He believes that the PCBs might have contributed to the early onset of the disease. A review of the relevant literature uncovers two animal studies that injected infant mice with very high doses of concentrated PCBs. A significant percentage of these mice developed cancers, though most were benign and none were small cell. There are also a couple of epidemiological studies, but they are at best equivocal and produced no statistically significant results. Finally, however, two well-respected doctors have examined Client and are willing to testify that based on their examinations, together with a review of the literature and the elimination of other possible causes, "PCBs more probably than not promoted Client's lung cancer."

How should you approach the case? First, it would be nice if more research could be done. But who is going to pay for it? Moreover, there is a statute of limitations period that requires the case to be filed within a set period of time following the first discovery of the possible link. This time is usually around two years and can be as short as six months. But even if there were no limitations period, Client might be dead before any research is completed. This would not terminate a lawsuit, but you would naturally prefer to bring the action while he was alive. The reality is that the suit must be brought now, under less than ideal circumstances.

From the legal system's perspective, science, though it might sometimes seem as swift as the Aesop fable's hare, can be very ponderous at times. It might take ten years or more to study the link between PCBs and lung cancer adequately, and by then, most of the litigation is done or will be on appeal. At the very least, a lot of decisions will have been made without the benefit of a fulsome research effort.

At the same time, however, if there are enough Clients out there, the issue will spur researchers to study the matter. This is what occurred with silicone implants. The tort litigation system provides ample incentives to find links between substances with real or contrived toxicity and defendants with "deep pockets" whose products contain those substances. The tort system effectively primes the pump to get research flowing. This can be a salutary aspect of the system. The problem is that the tort system's shut-off valve is not finely calibrated, so once research starts flowing, it can quickly turn into a flood. When this happens, funding resources are wasted investigating phantom links that are supported by little more than anecdotal reports and allegations in civil pleadings.

A more vexing aspect of data collection for the law comes from the basic incongruity between what scientists study and what the law is interested in knowing. For obvious reasons we cannot test the link between PCBs and lung cancer directly on humans. Animal studies and epidemiological studies are extraordinarily relevant, but they usually give us less confidence in the conclusion than we might desire. The law invariably relies on applied science, which typically involves extrapolating from controlled laboratory tests that are highly artificial or generalizing from noncontrolled field tests that contain many confounding variables. Applied science is hard to do, and the data usually are not sufficiently compelling to allow policy makers to sleep peacefully at night. Finally, science tends to study effects in populations, whereas at least at the trial court level in my example, the law is interested in effects on an individual. Science does not particularize with exactitude. Although the law and especially the civil litigation system also do not demand exactitude, the leaps required between population effects and the effect on some individual can be large and precarious.

Does Anybody Here Have a Slide Rule? Understanding the Science

Further complicating the law's use of science, legal consumers of scientific research often have little understanding of the product they are buying. In most areas of the law, those using science have little or no training in the subject. This is true for judges, jurors, legislators, and to a lesser extent, administrators. All judges and most legislators and administrators come from the ranks of lawyers. These are people who typically ended up in law school because their prospects in science and math were dim. Fewer than 10 percent of all students attending law school have undergraduate degrees in fields that require substantial math and science training, such as the natural sciences, math and statistics, computer science and engineering.[30] Not only do they not have training in the particular subject, they have a more profound disability: most lawyers have little or no appreciation for the scientific method and lack the ability to judge whether proffered research is good science, bad science, or science at all.

The American trial process adds a substantial wrinkle to the consumption of scientific information by dividing many of the decisions between two parties, the judge and the jury. Under the Constitution, people have a right to a jury in all criminal cases as well as in all federal civil trials in which "the value in controversy shall exceed twenty dollars." Most states also guarantee juries in civil cases. It is safe to say that if few judges are trained in rocket science, even fewer jurors have this training.

Compounding the lack of training in math and science is the method by which most science is introduced to the legal decision maker. For the juror, virtually all science is provided through oral presentations that might or might not be accompanied by overheads, slides, or other visual displays. They are virtually never given written reports. Similarly, although usually buttressed by written statements, legislators receive most of their information through oral testimony. Also, as anyone who has ever testified before a legislative committee can attest to, legislators constantly enter and exit hearings in order to attend to the press of other business. The complexities of science, however, are probably not best presented exclusively through oral testimony. While maybe less dramatic, most people comprehend complex information better when they have read it on the page, sometimes over and over again. The combination of oral and written presentation undoubtedly is the best for comprehension.

Compounding the problems still further, the trial system operates on an adversary model, and most experts who testify in legislative hearings are there because they have some ax to grind or someone's ox to gore. In practice, this means that much of the information that reaches the legal system does not represent the scientific field more generally. Very often scientists at the margins of their disciplines are chosen to make the presentation because they are willing to be more extreme in the proponent's favor and thus come across as more certain of their conclusions. This practice tends to suggest that the scientific fields are more divided and polarized than they are in fact. Also, experts are selected, though this tends to be less true in the legislative context, because of their ability to persuade, not because of their scientific accomplishments. The smoother, more practiced, expert is always preferred over the more capable one. To their general credit, or possibly to the credit of their egos, legislators prefer "big names." Many legislative hearings are populated by Nobel laureates and well-known academics. This celebrity mentality does have its drawbacks, since movie stars and citizens with heart-rending stories to tell also get prime-time attention.

Where Does This Piece Go? Integrating Science into Other Information

Most legal problems, like most problems in life, require an analysis of a variety of factors before deciding what should be done. Many of the factors are either impossible to compare or the scale on which they should be compared is not obvious. This is the proverbial apples and oranges problem. Of course, apples and oranges can be compared in a number of ways—such as size, price, weight, subjective preferences, and so on—but what scale to choose is not dictated by the fruit. The law presents many analogous situations, problems in which the fruit do not tell us how to choose among them.

Suppose, for example, the Army Corps of Engineers is considering building a reservoir outside Houston, Texas, that would occupy 19,700 acres. The project would serve four specific purposes: water supply, fish and wildlife enhancement, navigation, and recreation. At the same time, this project would disrupt existing wildlife and affect the salinity levels downstream in the Gulf of Mexico in a way likely to devastate the shrimp and oyster populations (because of the decreased amounts of fresh water flowing into the gulf). Should the corps build the reservoir? Doing a standard sort of cost-benefit analysis, the decision maker would want to compare the benefits of an adequate water supply, the enhanced fish and wildlife, better navigation, and more recreation against the costs to existing wildlife and the impact on the shrimp and oyster industry along the gulf. But each of these factors has both an empirical component and a normative component. The need we attach to the factor of "an adequate water supply," for instance, depends on long-range weather forecasts and thus the probability of drought occurring, together with the misery and dislocation that might be caused by droughts. And how do we compare the fact that the reservoir will displace one type of bird but replace it with another? And how is increased recreation evaluated? The loss of the shrimp and oyster industry is largely economic, though the value of these creatures to the way of life of the gulf communities should not be ignored. Would New Orleans be the same place without oysters?

In the realm of policy formation, the apples and oranges problem is considerably more complicated. It is more like comparing, on the one hand, apples, cherries, macadamia nuts, and aluminum foil with, on the other hand, oranges, peaches, peanuts, and string. Science can tell us such things as how mealy the apples are, the weight of the cherries, the juice content of the oranges, and how long the string is, but science cannot tell us how to combine this information to make a decision. The economic science of cost-benefit analysis can certainly offer some guidance on this matter, but it remains some distance from objectifying these tasks. The final policy choice between apples et al. and oranges et al. is likely to remain for some time a messy and intuitive subjective process of the human mind.

Does the Fork Go on the Left or the Right? Cultural Conflicts Between Law and Science

On the surface, at least, law and science appear to clash culturally in an assortment of ways because of their seemingly very different approaches to the world. Most scholars describe the subject of the law's use of science in these basic terms.[31] Science progresses while law builds slowly on precedent. Science assumes that humankind is determined by some combination of nature and nurture, while law assumes that humankind can transcend these influences and exercise free will. Science is a cooperative endeavor, while most legal institutions operate on an adversary model. Of course, these differences are largely superficial, and neither law nor science fits its respective stereotype fully. We know from Thomas Kuhn, for instance, that science too can be slow to throw away precedent.[32] Also, many areas of law, such as insanity, assume some measure of determinism. Scientists can be highly competitive, while lawyers have been known to cooperate. Law and science, therefore, do share some of the cultural dispositions of the other. There is no reason to believe that they could not yet reconcile their cultural peculiarities in a common goal.

The more substantial cultural clash harkens back to the hydra metaphor I suggested in Chapter I. And remember that there are three hydras. While I focus primarily on two of them, law and science, religion is walking right alongside them, both as a historical blood relation and as a continuing companion and competitor for attention. T. S. Eliot asked:

> Who is the third who walks always beside you?
> When I count, there are only you and I together
> But when I look ahead up the white road
> There is always another one walking beside you.[33]

These beasts compete with one another over the functions we traditionally associate with these three disciplines: descriptions of the natural world and the people who inhabit it, establishment of moral and ethical standards, and the regulation and enforcement of behavior. But law, science, and religion do not clash because they operate in different spheres doing different things. They clash because they operate in the same spheres doing the same things. Hence, although these three competing institutions surely have different cultures, they clash for the more interesting reason that they are involved in a struggle for power.

Like most struggles between warring parties, this one among law, science, and religion is not simple or straightforward. This struggle, as is true for most wars, contains many major battles, assorted minor skirmishes and treacherous intrigues behind the lines. It must be emphasized, however, that this war is being fought across the full societal spectrum, though my focus is trained on a small corner of the conflict. In this small sector, the terms of engagement have been mostly set by the lawyers, politicians and administrators. To press the metaphor, science effectively has been captured by the law and is in service to it.

III

THE GATEKEEPERS
Scientific Expert Testimony
in the Trial Process

*But there is one way in this country in which all men are created equal—
there is one human institution that makes a pauper the equal of a Rocke-
feller, the stupid man the equal of an Einstein, and the ignorant man the
equal of any college president. That institution, gentlemen, is a court. It
can be the Supreme Court of the United States or the humblest J.P. court
in the land, or this honorable court which you serve. Our courts have
their faults, as does any human institution, but in this country our courts
are the great levelers, and in our courts all men are created equal.*

—HARPER LEE, *To Kill a Mockingbird*

*If it had not been for these thing, I might have live out my life talking at
street corners to scorning men. I might have die, unmarked, unknown, a
failure. Now we are not a failure. This is our career and our triumph.
Never in our full life could we hope to do such work for tolerance, for
joostice, for man's onderstanding of man as now we do by accident.*

*Our words—our lives—our pains—nothing! The taking of our
lives—lives of a good shoemaker and a poor fish-peddler—all! That last
moment belongs to us—that agony is our triumph.*

—BARTOLOMEO VANZETTI

I n the spring of 1996, settlement negotiations continued over the final
amount Dow Corning and several other defendants should pay to settle claims that their silicone gel breast implants caused or might yet
cause the women who received them to become ill. These illnesses

ranged from relatively minor discomfort to life-threatening autoimmune disorders. Clinical observation, anecdote, and rumor, all multiplied by intense media coverage, seemed to point to a connection between the implants and illness. Some physicians, for instance, reported that their patients became ill soon after the implants were put in and that the patients' symptoms disappeared when the implants were removed. What began as a small wave soon grew to tidal proportions. The sum of the lawsuits exceeded many billions of dollars, brought by an estimated 500,000 plaintiffs. There remained no question whether to pay, only how much. The legal proceedings were nearing their end, yet the scientific proceedings had only just begun. A major epidemiological study published that same spring indicated that there was, at best, only a slight statistical association between the defendants' products and the more serious problems allegedly suffered by users of the implants. This study joined several others that had found no significant statistical association between breast implants and serious health problems. A Dow Corning official described the ordeal his company continued to endure as a "Kafkaesque nightmare."[1]

In the summer of 1996, after spending seven years in jail, George Franklin was set free. Franklin had been convicted of a murder that had occurred twenty-one years before, largely on the basis of the testimony of his daughter. His daughter testified that she had failed to come forward sooner because she had repressed the memory of the killing. Through therapy and, as was to be learned later, hypnosis, she was now able to recall specific details of that terrible day. The jury believed her and convicted him. After seven years of mainly lost legal battles, it turned out that she also "remembered" her father killing another person as well. Other evidence, however, unambiguously excluded Franklin as a suspect in that case. Based on the obvious fallibility of their primary witness, the state moved to drop all charges against Franklin. On the way out of the courtroom, Franklin's attorney told reporters that Franklin's ordeal had been "Kafkaesque."

Franz Kafka is invoked often in the face of legal procedures that appear to result in arbitrary outcomes. But Kafka's relevance to the silicone gel litigation and George Franklin's case is not entirely straightforward. Kafka, in his great and disturbing novel *The Trial*, related the story of K who, however much he persisted, was closed out of the legal process. In contrast, Dow Corning enjoyed unencumbered access to the courts, and George Franklin received a full trial and a complete panoply of appellate review. Where K's nightmare stemmed from too little access to the law, Dow Corning and George Franklin experienced the nightmare of too much legal process. Their nightmares are equally disturbing.

In K's futile search for justice, he repeatedly finds his access to the law blocked. Late in the novel, for instance, K meets a priest in a cathedral who describes the entranceway to the law through the story of another who sought to enter it. "Before the Law stands a doorkeeper," the priest explains. But the doorkeeper refuses to allow the man to enter and holds out little promise that he shall be permitted entrance later. The priest describes the man's frustration as he stands before the door:

> Since the door leading into the Law stands open as usual and the doorkeeper steps to one side, the man bends down to peer through the entrance. When the doorkeeper sees that, he laughs and says: "If you are so strongly tempted, try to get in without my permission. But note that I am powerful. And I am only the lowest doorkeeper. From hall to hall, keepers stand at every door, one more powerful than the other. And the sight of the third man is already more than even I can stand."[2]

The priest concludes by stating expressly the moral the man should draw from the story of the doorkeeper: "The law . . . should be accessible to every man and at all times."[3] Consistent with this lesson, fairness has long been thought to demand that every person deserves his—and all evidence deserves its—"day in court." More and more, however, this basic tenet is being challenged, especially regarding scientific evidence. As Dow Corning and George Franklin can attest, there are considerable costs associated with giving every person and every claimed scientific discovery a day in court. Admitting bad science poses similar risks to excluding good science. A doorkeeper who either excludes all who seek entrance to the law or admits them all fails to exercise the judgment necessary to do justice. Sadly, such doorkeepers are not solely the product of the dark imagination of one Czech novelist; they can be found at courthouse doors throughout the United States.

Guarding the Gate

The costs associated with having arbitrary doorkeepers have not gone unnoticed by the courts themselves. In 1993, the United States Supreme Court held that federal trial judges must be "gatekeepers" and ensure that the claimed scientific basis for expert testimony is valid. The decision, in the case of *Daubert* v. *Merrell Dow Pharmaceuticals, Inc.*, was widely heralded as a major transformation in the way federal courts (and since *Daubert*, most state courts) respond to scientific experts. As with Kafka's doorkeeper, the Court's gatekeeper is intended to stand guard over the processes of the law; unlike Kafka's doorkeeper, whose power is of an arbitrary and capricious sort, the *Daubert* gatekeeper is expected to bring reasoned principles to the task of deciding which scientists may enter. While the courts have yet to fully achieve this goal, it is clear that they are aware of their responsibility. Since 1993, the Supreme Court has taken up three scientific evidence cases, including *Daubert*.[4] This must be compared to the entire 204-year previous history of the Court in which it decided no such cases. The next several years will determine whether *Daubert* was an enlightened step forward in the way the law uses science or a stumble backward into the darkness of a "Kafkaesque nightmare."

The *Daubert* case began as a straightforward problem concerning the use of scientific expert testimony to prove causation in a civil litigation.[5] The lawsuit was brought on behalf of Jason Daubert, who suffered from a tragically common set of birth defects. The plaintiff claimed that his birth defects resulted from his mother's ingestion of Bendectin, a drug produced by Merrell Dow Pharmaceuticals that was widely prescribed for severe morning sickness throughout the 1960s and 1970s. The defendant, Merrell Dow, claimed that no competent scientific research indicated any causal relationship between Bendectin and birth defects. The trial court agreed with Merrell Dow and dismissed the suit before trial. On appeal, the Ninth Circuit Court of Appeals agreed with the trial court's summary disposition of the case. The plaintiff then appealed the case to the Supreme Court, and the Court agreed to hear it.

Before the Supreme Court, the plaintiff argued that the applicable rules of evidence established a limited role for judges in screening scientific evidence. It was the jury's primary responsibility to evaluate the science, and thus the case should have gone to trial. At trial, according to the plaintiff, the animal studies and epidemiological research would be sufficient to allow a jury to conclude both that Bendectin causes birth defects and that Jason's birth defects were caused by the drug; or at least that a reasonable jury could so find by a preponderance of the evidence. However, even the plaintiff recognized the weakness of the scientific claim. The research results were at best ambiguous and might more accurately be described as supporting the defendant's position. The plaintiff sought to buttress the basic research with a reanalysis of the several studies that had found little or no association between Bendectin and birth defects. Researchers used a "meta-analysis" to combine the slight but statistically insignificant effects found in the other studies and claimed to find statistically significant results. This research method is not entirely novel, but it contains enough dangers that it remains controversial, especially in specific applications. In addition, at the time, this work was tentative and had yet to be published in a peer-reviewed journal; in fact, to date, it still has not been published.

The defendant, Merrell Dow, argued that judges have a responsibility to screen all evidence to ensure that it is relevant and trustworthy. Merrell Dow emphasized the logical argument that invalid science cannot assist jurors and thus judges must initially determine the validity of the science. Bad science is irrelevant. Moreover, because most scientific expert testimony is complex and sometimes overwhelming, judges must be especially vigilant when assessing the basis for claimed expertise. The rules of evidence, according to Merrell Dow, treat scientific evidence cautiously and the experts who offer it with skepticism. Applying this rigorous standard should result in the exclusion of the plaintiff's experts and hence the collapse of the litigation.

The Supreme Court's opinion, written by Justice Blackmun, allowed both sides to declare victory. For the plaintiffs, and for plaintiffs' attorneys everywhere, the Court pointed out that the Federal Rules of Evidence were intended to liberalize the standards for admitting evidence. Thus, according to this principle, a jury ordinarily should be the arbiter of the weight of expert testimony. At the same time, for the defendants, the Court stressed that trial court judges must be "gatekeepers" who exclude scientific expertise not based on valid research. According to this principle, judges should play a substantial role in evaluating the value of scientific expert testimony. These seemingly contrary statements led to substantial initial confusion as to whether the *Daubert* test would be more "liberal" (i.e., result in more experts testifying) or more "conservative" (i.e., result in fewer experts testifying) than the test *Daubert* replaced. The truth lay somewhere in between.

Prior to *Daubert*, most courts, both federal and state, applied the "general acceptance" test, which was articulated in its modern form in the 1923 case of *Frye* v. *United States*. The general acceptance or *Frye* test called on judges to assess the degree of acceptance of a novel scientific technique in the field in which it belonged. The central premise of *Daubert*, in contrast, is that judges must assume the responsibility for conducting the evaluation of the scientific merit of expert testimony. Under a general acceptance test, the judge need not understand any of the science; he or she must merely identify the pertinent field in which the science falls and survey the opinions of scientists in that field. This nose-counting approach had the virtue of permitting judges to survey specialists who presumably knew the most about the science. It had the vice, however, of surveying the very people who had an interest in the outcome of the admissibility decision. Not surprisingly, if judges ask handwriting identification analysts whether their specialty of identifying handwriting is generally accepted, the answer is yes. This is not unlike asking tea-leaf readers whether tea-leaf reading is generally accepted. The general acceptance test thus depends on the scientific integrity of the field surveyed. Whereas particle physicists accept findings only after considerable rigorous research has been completed, many other ostensibly scientific fields accept findings more readily. This insight is the key to understanding the impact of *Daubert*.

Daubert requires judges to adopt a rigorous scientific mindset for evaluating the validity of the scientific research that supports expert testimony. The basic rule the Court adopted is that trial judges must assess the scientific foundation of expert testimony and be convinced that the science is more likely than not valid. In order to assist judges in this task, the Court suggested four factors that could be used, with others, to make the validity determination. These factors are whether (1) the hypothesis is testable and has been tested; (2) the error rate associated with the use of the science is not too great; (3) the basic research has been published in a peer-reviewed journal, and (4) the science supporting the expert opinion is generally accepted in the scientific community. In effect, the *Daubert* holding calls on judges to assess scientific research in much the same way scientists might. It incorporates the essential values of the scientific culture. Yet judges are not, nor should they be, as rigorous as particle physicists in evaluating research. At the same time, judges should be more rigorous than many scientists who, until now, have been permitted to testify routinely. Therefore, in practice, *Daubert* will be more liberal and lead to the admission of more expert testimony when it comes from fields with a long tradition of rigorous testing; but it should also be more conservative and lead to the exclusion of expert testimony when it is based on fields that do not have a tradition of rigor.

The *Daubert* litigation itself exemplifies this lesson. After the Supreme Court decided *Daubert*, it returned the case to the United States Court of Appeals for the Ninth Circuit to be resolved in accordance with the new standard. In a somewhat unusual move, the appellate court did not return the case to the trial court, instead finding that the science clearly did not support the plaintiff's claim and that no reasonable judge could find that it did. The Ninth Circuit interpreted the Supreme Court's new standard as a rigorous one and dismissed the plaintiff's claim. Judge Kozinski, a judge well known for his colorful opinions, explained, "Our task, then, is to analyze not what the experts say, but what basis they have for saying it."[6] He captured the spirit of the new standard when he observed that judges must "not ignore the fact that a scientist's normal workplace is in the lab or the field, not the courtroom or the lawyer's office."[7] This new responsibility, he stated, means that "federal judges ruling on the admissibility of expert scientific testimony face a far more complex and daunting task in a post-*Daubert* world than before."[8] The plaintiff again appealed, but this time the Supreme Court did not take the case.

Judge-Scientists?

A common response to the *Daubert* rule, a rule that requires judges to be sophisticated consumers of science, is that judges do not have the capacity for the job. In fact, in the concurring opinion in *Daubert* itself, Chief Justice Rehnquist warned that this new validity test would require judges to become "amateur scientists," a job for which they are ill suited. This complaint, however, is not easy to understand. First of all, it might very well be true that judges today are not well trained to evaluate science. But most judges are intelligent and well educated and there is no reason why they cannot with diligence master the basics of the scientific method and statistical theory. Second, in many areas, trial courts are the principal, if not sole, consumer of scientific expertise. Outside the courtroom there is a limited market for techniques such as polygraphs, bite-mark analysis, fiber analysis, fingerprints, DNA profiling, handwriting identification analysis, and so on. Since the courts are the primary consumers of this science, they should learn to evaluate what they are getting for their dollar. Finally, someone in the legal system must evaluate the merits of the science. Chief Justice Rehnquist's comment reveals little faith in judges' abilities to critically assess science. The obvious question is, if judges cannot assess this evidence, how can juries? And if juries cannot do it, then who will?

The Jury's Role

Inevitably, every discussion about the admissibility of evidence resolves into a debate about the jury system. The same is true for scientific evidence. The American trial system divides responsibility between the judge and the jury, and thus every task taken from one is assumed by the other. It is not entirely a zero-sum game, however, since increased judicial scrutiny of scientific evidence that is later admitted might nonetheless be rejected by the jury. But if increased scrutiny leads to exclusion, the jury will never hear it. As a general matter, then, increasing judicial control lessens the power of the jury. Yet, as a society, we are committed to finding justice through the wisdom and judgment of our neighbors in the community. Indeed, this is a widely cherished principle of our legal system. In order to allow jurors to make informed decisions, our rules of procedure and evidence must give them all or most of the facts.

Of course, this idealized version of the American jury system is just that, an ideal never fully or even largely achieved. Juries rarely resemble a cross-section of the community; the goal of most people called to jury service is finding a way to avoid it. In addition, given the nature of the adversarial system, only the most naive observer could believe that juries are privy to "all" the facts; instead, juries hear highly practiced alternative stories that only roughly approximate what might be termed reality.

The law has never given juries the free reign that myth assigns them or critics perpetually fear. The law of evidence is replete with privileges, ranging from attorney-client to physician-patient, and rules of exclusion, such as the hearsay rule and rules against use of certain kinds of character evidence, so juries are already precluded from hearing significant amounts of information. For example, in the prosecution of William Kennedy Smith for rape, the judge did not allow testimony that would have informed the jury that other women had had similar experiences with the defendant. At the time of his trial, the rules of evidence excluded this sort of prior bad act evidence on the basis that it would too greatly prejudice jurors by inviting them to punish the defendant for acts for which he was not on trial.[9] In addition, judges regularly set aside verdicts as being against the evidence or modify money awards they deem to be too excessive. In Stella Leibeck's suit against McDonald's, for instance, the jury originally awarded her $2.7 million in damages for injuries she received when she spilled a cup of scalding McDonald's coffee on her lap. Although it did not stem the criticism that followed the verdict, the judge reduced the award to $640,000. Juries clearly are an integral part of the American trial system. Just as clearly, juries are carefully controlled and monitored by judges to ensure that they act reasonably and in accord with the evidence.

In the area of scientific evidence, unencumbered access to the legal process produces costs as great as and sometimes greater than those restricted access produces. For example, in cases commonly referred to as "mass toxic torts," tens (and sometimes hundreds) of thousands of plaintiffs can be involved. The huge numbers of lawsuits associated with such products as asbestos, agent orange, Bendectin, silicone gel, and lead have transformed the civil justice system. In the silicone gel breast-implant cases alone, a stunning 500,000 plaintiffs claimed that their injuries and illnesses were attributable to the defendant's product. Whatever the "truth" behind these claims, simply permitting them to be litigated can bankrupt the defendants. *Daubert* instructs judges to exercise control over the entranceway because such great costs are at stake there. *Daubert's* ultimate legacy for the law is its insistence that legal policy and legal outcomes be crafted in light of a sophisticated appreciation for the complexity and subtlety of the state of the art of the science. This integration of law and science will not be simple and it will not be straightforward. Since neither law nor science is uncomplicated, few should expect their marriage to be.

Integrating Law and Science in the Courtroom

The legal system is engaged in a delicate balancing act when it comes to scientific evidence. Some of the difficulty stems from the inherent differences between the legal and scientific enterprises. Many legal scholars point out that science seeks truth while law seeks justice. This observation is much too simplistic, however, and manifests little appreciation for the culture of science and too facile an understanding of the culture of law. Truth, after all, should be no small component of justice. If silicone gel implants do not cause autoimmune disorders, then the $24 million that Pamela Johnson won in compensatory and punitive damages in 1994 against Dow Corning is hardly just or fair. A real and central difficulty in reconciling the trial process and scientific research is the very different time frames in which the two operate. Law must make decisions within a finite period of time, periods typically too short for substantial scientific work to be done—at least in the early cases. Science works with larger time frames, with major research programs spreading from years into decades. But this difference tends to mask an essential similarity: both science and law seek to make the best decision possible given the information available. To do so, they need to be fully informed of the facts and fully appreciate the values at stake in the decision. In fact, both science and law establish guidelines for avoiding factual errors that policy indicates are too costly.

The concept of error is explicit in the trial process. The legal system allocates the costs of errors through a variety of mechanisms, the most commonly known being burdens of proof.[10] In ordinary civil cases, in which society is largely ambivalent over who prevails and thus who over time bears the cost of mistakes, the plaintiff bears the slight burden of showing that his version is more likely than not true. This "preponderance of the evidence" standard operates largely as a tie-breaking mechanism and is placed on the party wishing to alter the status quo through litigation. Where the costs of error are much greater, as in criminal cases, the burden of proof increases proportionately. In criminal cases, the state must prove its case beyond a reasonable doubt. Society has chosen this high burden because the costs associated with incarcerating an innocent person are great. But the system does not demand certainty, and indeed it is well understood that sometimes guilty defendants go free and innocent defendants go to prison. Oliver Wendell Holmes estimated that the high burden of proof in criminal cases suggests that, as a society, we have decided that it is better to let ten guilty defendants go free than to convict one innocent person. Holmes might have overstated the ratio somewhat, but his point is important: more defendants can be convicted if we lower the standard of proof. But these greater numbers will include both the guilty and the innocent.

Science, too, is preoccupied with the costs of error. Most scientists begin with the assumption that the phenomenon they are studying does not cause the effect that they expect. In other words, the standard method of science is to presume "innocence" and only with strong proof reject that presumption. This presumption of innocence in science is called the null hypothesis. Over time, certain conventions have arisen in science regarding the strength of this presumption. Stated another way, a certain amount of evidence is needed before rejecting the null hypothesis and accepting the alternative hypothesis that the treatment variable caused a particular effect. The convention applied by most scientists is that the null hypothesis should not be rejected unless the chances of making a mistake are less than five in one hundred. This standard is referred to as a confidence level or p-value. This .05 rate, however, is merely a convention, and most scientists today take findings with rates between .01 and .10 very seriously. In fact, what standard a scientist adopts is ultimately dictated by policy, not science.

Consider, for example, a hypothetical drug called miracle-X. Let us suppose that two teams of scientists are studying this drug. One team is interested in the drug's effects on the AIDS virus, while the other is studying its effectiveness in smoothing wrinkles when applied topically. Each team conducts clinical trials and compares the effects of the drug to control groups that only receive placebos. Assuming that both teams find that the drug had positive effects, the question arises whether those effects are attributable to the drug or are just a matter of chance differences. In deciding whether to conclude that the drug had an effect, there are two possible errors the scientists might make. First, the scientists might conclude that the drug had an effect when it did not; this error is called a Type I error. Alternatively, the scientists might conclude the drug had no effect when it did; this error is called a Type II error.

Scientific convention, which is quite conservative in practice, holds that Type I errors should be made no more than five times out of one hundred; this is the .05 p-value. However, this convention is not divinely inspired. It is merely the scientists' way of being conservative regarding conclusions that claim an association between some cause and some effect. In the example, in fact, regarding the drug's effect on AIDS, we might be more concerned with making a Type II error—concluding that the drug has no effect when it does. Since AIDS leads to death when left untreated, researchers should be inclined to resolve uncertainty in favor of finding an effect and thus avoiding a Type II error. When it comes to wrinkle treatment, however, a conservative approach might very well be the wisest. Any side effects whatever should lead to substantial concern about making a Type I error.

When courts decide the admissibility of scientific evidence, they confront very similar policy decisions. Consider the issue presented in *Daubert* itself, whether Merrell Dow's drug Bendectin caused Jason Daubert's birth defects. In deciding the admissibility of scientific expert testimony, the court might make one of two types of error. It might admit the evidence when in fact Bendectin does not cause birth defects (what we might term a Type I error), or it might exclude the evidence when in fact Bendectin does cause birth defects (what we might term a Type II error). At the time the case was being litigated, the science was somewhat uncertain. The court, therefore, was confronted with the costs of making a mistake under conditions of uncertainty. If the court admits the evidence when it has no ill effects, it potentially drives a useful drug and possibly the drug's manufacturer from the market. The fact that jurors also evaluate the validity of the science does not affect this risk; the statistical risk of error is inherent in the scientific enterprise itself and is not correctable by independent review, even by the most sophisticated reviewers. If the court excludes the evidence when the drug has ill effects, it deprives the plaintiffs from being compensated for injuries caused by the defendant. Even if the science develops sufficiently someday to demonstrate a relationship, it is too late for those early plaintiffs; jurisprudential considerations of fairness preclude plaintiffs from relitigating their claims when better science becomes available. Whether to err on the side of the plaintiff or the side of the defendant in a particular case is thus purely a policy judgment.

Unfortunately, the scientist's concept of statistical error does not translate directly into the judge's concept of legal error. We cannot say, therefore, that a study that is statistically significant at the .05 level of confidence will lead judges, if they admit the evidence, to make only five errors (of the Type I variety) out of one hundred. Hence, there is no true correspondence between statistical confidence and legal burdens of proof. The confidence level is merely a statistical statement, and it does not incorporate a host of factors that should affect a judge's decision whether to admit evidence. In particular, the many errors that might be made in research methodology must be taken into account by judges. For instance, research examining possible health hazards associated with silicone gel breast implants that only studied women who had received implants following a mastectomy might contain statistically significant findings that the implants are associated with serious illness. But generalizing this work to women who received their implants for cosmetic reasons alone would be a mistake. Women who acquire implants after a mastectomy are differently situated than other women, both in the procedure used to insert the implant and in their health histories.

Another limitation to moving from statistical significance to legal significance is the fact that most scientific research examines the general relationship between variables. Trial courts are usually concerned with specific effects on specific individuals. While science attempts to discover the universals hiding among the particulars, trial courts attempt to discover the particulars hiding among the universals. We have very good research, for example, that cigarette smoking causes lung cancer, and this research easily meets conventional standards of statistical significance. But the ultimate legal issue in a particular suit against a cigarette manufacturer concerns whether the plaintiff's lung cancer was caused by the defendant's cigarettes. It is this fact that the jury must find by a preponderance of the evidence. The state of the art of the science is not as advanced in determining whether a particular plaintiff's lung cancer is attributable to the defendant's cigarettes as it is in establishing the general association of cigarettes and lung cancer. Because other products in today's society are associated with lung cancer, it would be a profound mistake to glance at the .05 confidence level and say that such a conservative estimate meets the preponderance standard (which, by definition, is merely greater than 50 percent). The statistical confidence level is merely one factor among many that contribute to a judge's or jury's confidence that the plaintiff has met or failed to meet the burden of proof. Statistical significance is clearly an important aspect of determining whether plaintiffs have met their burden of proof; but it will never alone be sufficient.

The integration of science into the law is not just a matter of assessing the strength of the science; it also requires an evaluation of the needs of the law. Specifically, judges must assess the legal or practical significance of scientific research. The same science might be very valuable in certain cases and entirely irrelevant in others. Research results having statistical significance might very well have no practical significance. Courts should find that different legal contexts sometimes lead to very different judgments regarding the usefulness of scientific evidence. Consider, for example, psychiatric predictions of violence. A person's propensity toward violence is considered relevant in a wide variety of legal contexts, ranging from parole hearings to capital sentencing hearings. Psychiatric predictions are probably most frequently used in proceedings for involuntary commitment to mental hospitals. Research indicates, however, that psychiatric predictions of violence are not very good. A legal decision maker's determination whether to rely on a prediction of violence should depend substantially on what costs are associated with making a mistake. A prediction of violence in the parole context might be relied on more than one in the capital sentencing context because the costs of making a mistake are so much less. The validity of the science, here psychiatry/psychology, is largely the same in the two contexts. It is the legal context that has changed and that thus affects the utility of the science.

When courts use science, they seek to fit what is known about legally relevant facts into a pre-existing legal fabric. The legal fabric is defined by value-based considerations. Consider, for example, the admissibility of the so-called "black rage" defense. This was the theory William Kunstler proposed when he represented Colin Ferguson against ninety-three counts of murder, attempted murder, assault, weapons charges, and civil rights violations arising out of Ferguson's shooting spree on a commuter train of the Long Island Railroad in December 1993. Although the defense was dropped after Ferguson fired Kunstler and began representing himself, it is likely to return in other cases, and it closely resembles an assortment of other defenses that have been catalogued together as "abuse excuses."[11] Some of these include battered woman syndrome, battered child syndrome, posttraumatic stress disorder, and the "twinkie defense." The basic insight of black rage defense is the same as that of most of the syndrome-styled defenses: these defendants' experiences so influenced, if not caused, them to kill that they now cannot be held accountable for their actions. The viability of these defenses depends on two ostensibly separate conditions being met. First, the law must recognize that certain psychological states excuse or justify conduct that would otherwise be criminal. Second, assuming the first is met, scientific research must indicate that people experiencing what the defendant experienced suffer the psychological state that the law recognizes as an excuse or justification. Politics drives the first condition and science should drive the second. With most of these syndromes, however, politics drives both conditions.

When Science Is Politics

"If the law supposes that," said Mr. Bumble . . . "the law is a ass — a idiot."

— CHARLES DICKENS

The courts' move toward a sophisticated integration of legal policy and empirical science has been slowed by the impurity of much of the science they see. A large element of politics drives many ostensibly scientific findings. The quality of much of the science that is urged on courts is, to put it mildly, weak. In fact, it is not science at all. Many claims to science are really assertions of policy wrapped in the guise of science. If treated as bona fide science, this masquerade distorts legal outcomes. Until courts understand the scientific method, they will be unable to distinguish good science from the sort of pseudoscience that is forwarded to advance political agendas. This policy aspect of science has been especially marked in the social sciences, where the law has confronted an excess of syndrome-styled testimony. Indeed, many lawyers seem to suffer from "syndromic lawyer syndrome," a pathological acceptance of simplistic explanations for complex human behavior that supports otherwise desirable legal outcomes.

Advocates for a variety of causes find it politically expedient to peddle their causes through expert testimony rather than through substantive legal change. Like Odysseus' Trojan Horse, experts can be a far more effective strategy than a frontal assault. But there are many dangers associated with this approach. Foremost, it undermines the legitimacy of the law by basing legal outcomes on the false premises of pseudoscience. In addition, the scientific explanation rarely encompasses the full political agenda, and thus reforms typically fall well short of proponents' desires. Finally, and perhaps of greatest concern for proponents, peddling policy as science permits opponents to corrupt the "science" to their own political objectives.

Two examples particularly well illustrate the phenomenon of politics masquerading as science. The rape trauma syndrome and the battered woman syndrome were each introduced as ostensibly scientific findings, and both nominally serve the political objectives of readily identifiable constituencies. In fact, they serve political objectives with which I personally agree. The rape trauma syndrome was originally embraced by prosecutors in order to increase conviction rates in cases notoriously difficult to win. The theory and research of the rape trauma syndrome, as originally offered by the psychologists Burgess and Holmstrom, was that women who were raped experience identifiable psychological responses, some of which are inconsistent with stereotypical expectations.[12] Hence, prosecutors would use expert testimony indicating that the woman "suffered" from the rape trauma syndrome in order to rebut claims that the alleged victim acted in ways inconsistent with having been raped and, more generally, to buttress her testimony when the defendant claimed consent. Courts generally permit this syndrome evidence in order to rebut assertions of inconsistency but are reluctant to allow it to show that the alleged victim was raped. No court allows it to prove that the defendant was the perpetrator.

The battered woman syndrome is offered to explain to jurors why a woman in a long-term violent relationship would not leave the violence and, moreover, why she might kill under circumstances that do not mirror traditional notions of self-defense. In many of these cases, the woman kills the batterer when he is sleeping or during another kind of lull in the violence. Traditional self-defense doctrine, however, requires that deadly force be used only in response to repel an *imminent* attack that might result in serious physical harm or death. Syndrome advocates contend that battered women become "learned helpless" and are thus unable to escape the violence. Also, according to the theory, they experience the danger as "imminent," despite appearances to the contrary, because of the cyclical nature of the violence. Just as with the rape trauma syndrome, therefore, the battered woman syndrome provides psychological explanations that conveniently satisfy already existing rules. Without the psychology, the rules would sometimes permit results that many syndrome advocates consider to be politically unpalatable. The battered woman syndrome has met with great success and is admitted in self-defense cases in the vast majority of states.

Despite their enormous success, these two syndromes are supported by a paucity of research. Although I concentrate my energies on the law's experience with the battered woman syndrome, the lessons drawn from this experience apply generally to the many areas of expert testimony that are based on good intentions but bad science.

The amount of research that actually exists to back the claim of expert witnesses who testify on the battered woman syndrome makes cold fusion look as solid as the second law of thermodynamics. The working hypothesis of the battered woman syndrome was first introduced in a 1979 book by Lenore Walker.[13] At the start, however, the hypothesis had little more support than the clinical impressions of a single researcher. Five years later, Walker published a second book that promised a more thorough investigation of the hypothesis.[14] In fact, however, this book was little more than a patchwork of pseudoscientific methods employed to confirm a hypothesis that the researchers never seriously doubted.[15] Indeed, the 1984 book would provide an excellent case study for psychology graduate students on how *not* to do empirical research. Yet, either because they shared the researchers' political agenda or did not look at or understand the science, judges welcomed the battered woman syndrome into their courts. Increasingly, however, legal commentators are realizing that this original conception was without empirical foundation and, perhaps more troubling, inimical to the political ideology originally supporting it. In short, in the law's effort to use science to make good policy, it is now obvious that the battered woman syndrome provides neither good science nor good policy.

Because I have often been misunderstood on this point, I must emphasize that I share most of the political values that drive battered woman syndrome researchers. Moreover, the battered woman syndrome, as originally promulgated in the language of self-defense doctrine, advanced several laudable goals. Foremost, it helped bring to the public's attention the horrifying pervasiveness of domestic violence. By highlighting the issue in the stark context of the criminal law, it contributed to concerted efforts to remedy the problem. Since the time the battered woman syndrome was first hypothesized, resources for victims of domestic violence have dramatically increased. Syndrome advocates deserve substantial credit for this improved situation. In addition, the battered woman syndrome highlighted the weaknesses inherent in the traditional conception of self-defense. The law of self-defense is largely driven by male conceptions of violence. Hence, in most jurisdictions, the defendant must show that she used a proportional amount of force and only to respond to an imminent harm. This is an idealized version of the way men fight. But a woman who is physically smaller than a man must defend herself in different ways. Whatever the validity of the notion of a "fair fight" for men, it cannot serve as a model of fairness for women. Finally, syndrome advocates deserve substantial credit for focusing researchers on the psychological dynamics of violence in intimate relationships. Although the law has become fixated on the syndrome model, many researchers are now conducting excellent, sustained research on the psychology of both battered women and the men who batter them. Some of this work is relevant to legal decision making, and future work will undoubtedly provide substantial insights into the psychology of domestic violence.

The battered woman syndrome ultimately fails because it was never a matter of science to begin with, yet it was treated as a "scientific fact" by courts. Good science does not serve the specific interests of any political viewpoint. Science has political consequences, but its results should be as free of political influence as possible. Courts began to accept the battered woman syndrome as such a fact, entirely failing to see its political basis. The scientific jargon of syndrome advocates was taken seriously by judges, with ramifications not altogether salutary for battered women. Judges refer to women "suffering" from the syndrome as "learned helpless," and, possibly, having fallen prey to an identifiable psychological disability. Courts describe women who kill after being subjected to years of abuse, who kill after pursuing every reasonable alternative, as psychologically disabled and as deserving to be excused for their action. The pathology of the violent relationship has become the pathology of the battered woman. To be sure, there are empirical realities that battered women face, and few critics of the syndrome would wish to keep this information from reaching the jury. Battered women continue to confront enormous societal obstacles, not the least of which include the police and legal system's inability to respond effectively to domestic violence. Also, the facts indicate that many battered women do not leave violent relationships, most often due to threats or lack of economic or social resources. All this contextual information is highly relevant to the reasonableness of a battered woman's decision to kill her batterer. Describing the situation as a syndrome, however, does not improve matters and in fact may make them much worse.

Many feminist scholars agree with the scientific critique on other than scientific grounds. These scholars criticize the demeaning connotations of syndrome-styled excuses for actions they consider to be reasonable and justified acts of self-defense. According to this view, the language of psychological disability rings too close to the archaic characterizations of women that for too long have dominated the law. The challenge that many feminist scholars have successfully assumed is explaining why traditional legal standards fail to account for the battered woman's situation. They reject the pathology model for the defense of battered women. It is not the battered woman who is mentally ill; the sickness lies with the batterer and the society that fails to redress the violence. Battered women who kill have either responded reasonably, and thus justifiably, or not. Continuing to treat battered women as less than rational actors demeans women, distorts science, and fails to do justice.

In fact, advocates of battered women who cast their arguments in the aura of science have sacrificed their politics and created a weapon for their political opponents. The syndrome casts the focus on the battered woman, rather than where the true pathology resides, with the batterer and society's too often ineffective response to the violence. In addition, because the syndrome describes the woman as suffering a psychological malady, prosecutors increasingly subject battered woman defendants to a searching psychological inquiry by the state's psychiatrists, thus victimizing many battered women still further. Moreover, because the syndrome has no empirical basis and thus might apply in any case, many defendants claim it when the facts of their cases fall well outside any reasonable conception of self-defense. For example, Lenore Walker sought to testify when a defendant hired a hit man for $10,000 to kill her husband.[16] These abuses cast long shadows and undermine the effort to make the law more responsive to the true challenges of domestic violence. Finally, because the battered woman syndrome is so empirically bankrupt, the courts have had to bend the rules of admissibility in order to allow it. But these rules apply to both sides equally. Therefore, prosecutors are given great leeway in introducing psychiatric theories that are about as well validated as the battered woman syndrome but that are not particularly friendly to the battered woman defendant.

Perhaps the most dramatic and disturbing example of opponents using the other side's politically based science against them comes from experience with the rape trauma syndrome. As noted, the rape trauma syndrome was borrowed by prosecutors from its noncontroversial therapeutic genesis in order to buttress a rape complainant's testimony at trial. In many rape prosecutions, especially those in which the defendant claims consent, the case comes down to who the jury believes, the defendant or the alleged victim. In fact, it used to be a regular practice among defense attorneys to introduce the prior sexual history of the alleged victim in order to discredit her. In the 1970s, however, virtually all jurisdictions established rules, called rape shield statutes, that strictly limit this sort of searching examination of the alleged victim's past. This reform was intended to protect against the victimization caused by a searching cross-examination into the personal details of the witness's life. It is thus ironic and profoundly unfortunate that defense attorneys' use of the rape trauma syndrome threatens this reform.

The theory of the rape trauma syndrome is that women who have been raped are more likely to exhibit particular psychological reactions. Increasingly, these reactions are associated with posttraumatic stress disorder. Defense counsel, on the other hand, would not be unreasonable in assuming that testimony that the alleged victim does not suffer from this syndrome might be used to show the defendant's innocence. The courts cannot very well allow prosecutors to employ psychological expertise and deny it to defendants. Even in an age of increased awareness of "victims' rights," the Constitution and the rules of evidence give defendants substantial procedural guarantees and, more generally, every benefit of the doubt. The problem, of course, is that the rape trauma syndrome itself is so elastic that usually the "right" expert can find the "right" result, depending on who is paying the fee.

More troubling, if the witness is found to suffer from posttraumatic stress disorder, a defendant can seek to explore her past history in agonizing detail in order to determine if there might be another explanation for the disorder. This is the sort of cross-examination by character assassination that was outlawed by rape shield statutes. The result is that many rape cases have become battles of the experts. The rape trauma syndrome, as policy-based pseudoscience is prone, has been manipulated easily to serve exactly the opposite end to that for which it was designed.

Witnesses are subjected to the abuses of the past through defense use of a modern innovation designed by those most sympathetic to the rape victim's plight. This should be a lesson to the next generation of syndrome advocates: science that is a matter of political convenience can be convenient for your political enemies.

Bad Science and No Science

Lord Chesterfield had this advice for his son: "To know a little of anything gives neither satisfaction nor credit, but often brings disgrace or ridicule."[17] At the same time, however, science often offers only a little knowledge on questions of legal significance or can offer more only when it is too late. Thomas Huxley, a scientist through and through, warned, "If a little knowledge is dangerous, where is the man who has so much as to be out of danger?"[18] The trial process must maneuver between these two rocky shoals: little knowledge of the pseudoscientist as a dangerous thing and dangerously unrealistic expectations of a scientific answer.

The social sciences and liberal political causes are not alone in raising ignorance of the scientific method to great jurisprudential and commercial success. In fact, probably the worst uses of courtroom pseudoscience can be found in the service of prosecutors in the field known as forensic science. The word forensic refers simply to the application of a particular discipline to the courtroom; the word science, among forensic scientists, appears to have widely varying meanings.

There are literally dozens of specialties in forensic science. Some of the better known are fingerprinting, DNA profiling, handwriting identification analysis, bite mark identification, fiber analysis, hair analysis, tool mark identification, and fire and explosion analysis. Some of the lesser known include gas chromatography, voice spectrography, manganese encephalopathy, hedonic damages, and many others with similarly impressive-sounding names. Unfortunately, neither their notoriety nor their impressive labels guarantee their value to the law.

An initial difficulty with the forensic sciences is the wide variability between the many disciplines that fall under this category in terms of the methods they use to validate their opinions and the qualifications necessary to practice their specialty. Some who work in these disciplines exemplify how the scientific method can be used in the service of the law, and some are little more than artisans who ardently believe in their art but would not know where to begin to test the validity of their conclusions. Undoubtedly, some are charlatans who, for fame, fortune, or ideals, don scientific robes to better peddle their opinions.

After *Daubert*, a critical issue that arose concerned the scope of the gatekeeping function. Since *Daubert* involved scientific expert testimony, many experts with less than exemplary scientific credentials or data began to renounce the science label and call themselves "specialists." For forensic "scientists" and others with little or no research to support their claims to expertise, the key to continued admission appeared to be to claim that *Daubert* did not apply to them.

Despite the obvious duplicity of this rejection of the mantle of science, there is some merit to it. The rules of evidence do not require that all expert testimony be scientific or that all experts be scientists. In fact, nonscientists, ranging from accountants to auto mechanics, testify routinely. Virtually all codes of evidence, federal and state, permit anyone with "scientific, technical, or other specialized knowledge," whether gained through education or experience, to testify. The question, then, is why should social and forensic scientists be held to a higher standard than auto mechanics? That many experts and their lawyer sponsors attempted this end run around *Daubert* is not terribly surprising. What is surprising is that so many courts bought this theory. Accordingly, for many courts the "critical question" after *Daubert* became how to distinguish between testimony that is "scientific" and testimony that is "nonscientific."

In the case of *United States* v. *Starzecpyzel*,[19] for example, the prosecution made the argument at trial that its handwriting identification analyst, a forensic scientist, should not be held to the high standards of science. As discussed in detail in Chapter I, handwriting identification analysts are not trained as scientists, nor do they understand the scientific method. In the immortal words of the Wizard of Oz, "Pay no attention to that man behind the curtain."

The judge in *Starzecpyzel*, Judge McKenna, agreed that the science of handwriting identification had yet to be validated. He observed that no research existed that substantiated the claims of handwriting experts that they could validly identify the author of a writing sample. In fact, one study investigating this question obtained fascinating results. The researchers found that a group of handwriting experts who were given a known sample and an unidentified sample and asked to compare them were 100 percent consistent among themselves, or, in scientific terms, the analysts demonstrated 100 percent reliability. However, the experts had 0 percent validity; they were all wrong. After reviewing the scientific support for handwriting identification expertise, Judge McKenna concluded that this forensic science was no science at all.[20]

Despite this conclusion, or perhaps because of it, Judge McKenna then found that handwriting expertise might nonetheless assist the trier of fact as specialized knowledge. Besides, he observed, jurors could compare the writing for themselves. In effect, Judge McKenna concluded that handwriting experts were allowed both because they could assist jurors and because jurors did not need them. This might be termed the chicken soup conclusion. Even though there is no evidence that it treats colds, it can't hurt. In the law, however, there are bones in the soup.

In explaining his decision, Judge McKenna likened the discipline of handwriting identification to that of a harbor pilot. If the law would permit a harbor pilot to testify by virtue of his experience, why not a handwriting expert? But implicit in the assumption that the harbor pilot gets to testify is the premise that a harbor pilot is not prone to ground ships on sandbars. The harbor pilot's knowledge has been tested the way all harbor pilots are tested, by experience. The same is not true for handwriting experts. Perhaps such experts have anecdotal support. A surplus of anecdotes amounts to little more than a surplus of anecdotes. After all, there is more than sufficient anecdotal evidence for both the "flat earth" theory and the "earth is the center of the universe" theory.

What, then, is science, and how can courts know it when they see it? And is this a question with which courts should be struggling, given that philosophers of science have been unable to give a definitive answer to it?

Several courts have relied on an example first described in *Berry* v. *City of Detroit*.[21] In this case, Doris Berry sued the City of Detroit for damages after a Detroit policeman shot and killed her son, Lee Berry, Jr., following a traffic stop. At trial, she introduced an "expert" to testify on the disciplinary procedures of the Detroit Police Department and their possible link to the shooting of Lee. Specifically, Berry's expert claimed that Detroit's loose disciplinary procedures had led to the shooting, thus making Detroit liable for damages. The expert had a master's degree in sociology and had some general experience as a police officer and in managing and training sheriffs' departments. He did not claim to be a scientist, nor had he studied the subject on which he testified — the relationship between police disciplinary policies and police officer behavior — scientifically.

In *Berry*, the Sixth Circuit Court of Appeals distinguished between scientific and nonscientific testimony as follows:

> If one wanted to explain to a jury how a bumblebee is able to fly, an aeronautical engineer might be a helpful witness. Since flight principles have some universality, the expert could apply general principles to the case of the bumblebee.
>
> On the other hand, if one wanted to prove that bumblebees always take off into the wind, a beekeeper with no scientific training at all would be an

acceptable expert witness if a proper foundation were laid for his conclusions. The foundation would not relate to his formal training, but to his first-hand observations. In other words, the beekeeper does not know any more about flight principles than the jurors, but he has seen a lot more bumblebees than they have.

It was an unfortunate aside because several other courts relied on this language to dismiss trial judges from their gatekeeping duties and to free experts of any responsibility to test their hypotheses scientifically. What might be termed the bumblebee criterion would seem to permit some witnesses to qualify as experts based on their experience with the phenomenon alone. This would have spelled the doom of *Daubert*, for virtually all failed science began or continues as the product of observation or experience.

Among the cases relying on the bumblebee criterion was one from the Eleventh Circuit, *Kumho Tire Co. v. Carmichael*. In fact, the Eleventh Circuit used the bumblebee criterion to reach the remarkable conclusion that an engineer's testimony regarding the cause of the plaintiff's tire failure was not a subject of "science."[22] Since engineers are applied scientists, this approach carved a fairly large loophole in the *Daubert* test. The law, in fact, relies almost exclusively on *applied* science, so this approach would have potentially exempted all experts. Indeed, the bumblebee criterion manifests a rather primitive understanding of the subject of science. Certainly observing bumblebees could be the subject of science. And it is simple enough to see why it might be a good idea to study bumblebee flight direction scientifically if it were somehow important enough to a court.

Suppose a bumblebee observer finds that bees "always" take off into the wind. Is this the sort of expertise that the law should depend on? It might be that the bees actually always take off toward a food source that they pick up with the wind. Or perhaps the bees were observed on the west coast where the wind is predominantly from the west. The bees don't take off into the wind, they take off in a westerly direction. The whole point of science is to test observation systematically. Observation provides a little bit of knowledge, but is it enough when the scientific method might be brought to bear and produce so much more?

In a unanimous decision, the Supreme Court in 1999 put an end to this debate.[23] Using the Eleventh Circuit case of *Kumho*, the Court held unequivocally that all expert testimony falls within the purview of *Daubert's* gatekeeping requirement for federal trial courts: "We conclude that *Daubert's* general holding—setting forth the trial judge's general 'gatekeeping' obligation—applies not only to testimony based on 'scientific' knowledge, but also to testimony based on 'technical' and 'other specialized' knowledge."[24] Significantly, the Court rejected the distinction lower courts had drawn between "scientific" and "nonscientific," with the former getting close scrutiny and the latter getting a free ride. Initially, the Court pointed out, the rule itself (Rule 702 of the Federal Rules of Evidence) makes no such distinction. *Daubert* had limited its discussion to scientific knowledge because "that [was] the nature of the expertise at issue" there.[25] Additionally, since all experts are granted great leeway in testifying to opinions, they all present dangers that must be checked through the gatekeeping function. Finally, the Court pointed out, "it would prove difficult, if not impossible, for judges to administer evidentiary rules under which a gatekeeping obligation depended upon a distinction between 'scientific' knowledge and 'technical' or 'other specialized' knowledge. There is no clear line that divides the one from the others."[26] In conclusion, the Court stated that, regarding all expert testimony, trial court judges "must determine whether the testimony has 'a reliable basis in the knowledge and experience of [the relevant] discipline.'"[27]

This brings us to the crux of the matter. The mistake the Eleventh Circuit made was to attempt to establish admissibility rules based on the nature of the testimony (is it science based?) rather than the nature of the legal question to be answered by the testimony. The court seemed to be driven by the concern voiced by many courts and commentators that the aura of a scientist's testimony is more likely to overwhelm jurors than that of a "mere specialist" or a "mere technician." But overwhelming the jury is not the only danger associated with expert testimony. There is a score of other concerns associated with experts who lack a reliable basis for their opinion, ranging from their introducing evidence that is otherwise inadmissible to their prolonging litigation and wasting time and resources. Moreover, although it is possible that some scientists cast a spell of certainty, many other experts might do the same. If jurors are so susceptible to the "scientist" label, then they are likely to be similarly affected by the "expert" label.

To the extent that nonscientist experts claim expertise on the basis of experience, their helpfulness to the law is not obvious. Our senses often prove dull. The law must be cautious before relying on experience in light of the relatively straightforward lesson of history that common sense about the empirical world—even that based on extensive experience—is often wrong.

The issue for the law of evidence, therefore, is whether a beekeeper is *good enough* in light of the law's needs as well as what it should expect to receive. This is a policy judgment. In *Daubert* itself, though it used a fanciful example, the Court recognized that underlying the admissibility decision lies the policy judgment of how demanding courts should be regarding the level of experience or the amount of research that is necessary before testimony is allowed. "The study of the phases of the moon," the Court suggested,

> may provide valid scientific "knowledge" about whether a certain night was dark, and if darkness is a fact in issue, the knowledge will assist the trier of fact. However (absent creditable grounds supporting such a link), evidence that the moon was full on a certain night will not assist the trier of fact in determining whether an individual was unusually likely to have behaved irrationally on that night.[28]

For good reason, the Court assumed that an expert could be found to testify that a full moon adversely affected a particular individual. Undoubtedly, such testimony would be based on "clinical experience." But it would not be "creditable." Experience in this case should not be enough, even if the expert disavowed any pretensions to testifying about "scientific knowledge."

Yet experience sometimes is enough, particularly when no more is available and no more can be expected. Because the law must be practical, it sometimes must settle for experience. Moreover, and again as part of its practical orientation, the law can rely on experience when there are good reasons to think it wasteful to demand more. Hence, a court might not permit a medical doctor to testify that, in his experience, drug X causes cancer, because it expects such opinion to be backed by creditable research. The court, however, might permit a surgeon to testify, based on experience, that a particular clamp does not leak when used in normal practice. It is not that this fact is not subject to general testing, since it might be that the doctor has not used the clamp in a wide variety of settings or is biased in his or her expectations. Instead, the court is making the policy judgment that the law demands no more. Auto mechanics, accountants, electricians, and plumbers testify routinely pursuant to this implicit judgment.

This approach has the advantage of also addressing the difficulty concerning the reach of *Daubert* which the Court addressed in *Kumho*. Distinguishing between science and nonscience for evidentiary purposes would have raised the specter that all ostensibly scientific experts who failed to test their hypotheses could still qualify as experts based on experience. Even more disturbing, it suggested that the best way to continue to testify would be to refrain from conducting any research at all, for that would suggest that *Daubert* might be applicable. Perhaps the most profound lesson of *Daubert* has been its effect on the professional fields that supply the legions of experts to the law. Whatever its effect on the law might someday be, it has already fundamentally altered the practices of the forensic and social sciences. Practitioners in these fields who make their living by testifying are preoccupied with the decision and have begun to respond to its dictates. Although it remains too early to say for sure, *Daubert* appears to have initiated a scientific revolution of sorts in these fields. The law is a consumer that receives only as good as it demands. Admissibility rules have a significant impact on the way and whether research is done. The law must tailor its admissibility rules to the needs it has for creditable expert evidence.

A Market Theory of Science

One of the signature aspects of science is that it "progresses." Although the idea that science progresses has taken some philosophical hits in these postmodern times, science's "forward" movement cannot seriously be doubted. I do not mean to suggest, however, that "progress" always means improvement. Thus, for example, development of the atomic bomb was certainly a scientific and technical advance of profound proportions; I would hesitate to say, however, that splitting the atom improved our lives. "Progress" is thus a descriptive term, whereas "improvement" is a value-laden or normative one. In contrast to science, to the extent law progresses, it is in a nonlinear and haphazard fashion that defies simple description. While I am fairly confident that Madisonian democracy is progress over Hobbesian monarchy, I would prefer not to have to prove it. The law is a little like a biological system. It evolves, certainly, but it is a basic error to suggest that it is progressing to some surpassing goal. In comparison, however arbitrary, science has the self-contained objective of describing, predicting, or controlling the natural world. It progresses to the extent that it does this better.

Why science does progress makes for fascinating sociological inquiry. The principal reason probably lies in some general market explanation. Foremost, science and especially its product, technology, create wealth. Moreover, in most circumstances, the amount of wealth depends on the success or validity of the principles or technology produced. With certain exceptions, the better mousetrap rewards its inventor and becomes a bestseller. Most scientific communities similarly reward recognized innovators, though the pecuniary rewards tend to be less generous. Hence, Einstein's theories largely replaced Newton's because they worked better. But the market theory operates on the assumption that the consumers care about the quality of the product they receive.

Yet the marketplace does not always work as advertised. Scientists sometimes fail to appreciate new theories, because of either self-interest or simply limited vision. Radical new scientific theories threaten the status quo, which by definition includes the audience for these theories. Scientists would not be human if they did not let their egos interfere with their acceptance of new work. In fact, what is most surprising is how well the community of scientists has subverted ego for truth. This might be the single greatest attribute of the scientific method. In the sixteenth century the church might have been able to dictate truth through coercion. Today, truth is power, and the powerful ignore this lesson at their peril.

Another even more substantial threat to the automatic adoption of the better mousetrap lies in traditional market failure. With technology in particular, an entrenched or better-situated mousetrap sometimes prevails against even a superior alternative. The example invariably cited to illustrate this phenomenon is the typewriter keyboard, often referred to as the Qwerty effect. Although the Qwerty effect has been attacked by economists as "myth,"[29] even as just a story it makes an important point. The Qwerty effect, a name taken from the top line of letters of the keyboard, occurs when a technology succeeds because of familiarity rather than superiority. The Qwerty keyboard was invented by Christopher Latham Sholes to avoid the repeated jams that occurred with the alphabetic layout. Contrary to popular belief, Sholes designed the Qwerty to speed up typing, not slow it down. Qwerty is outdated, since modern electronic keyboards do not jam like their mechanical predecessors. The awkward design also contributes to carpal tunnel syndrome and repetitive strain injuries. Nonetheless, because of familiarity with the old and the cost of replacing Qwerty with something new, it remains the default keyboard. Much of the science used by the law suffers from the Qwerty effect. In today's courtroom, however, it might be termed the Qwerty syndrome.

But while some products are obvious successes or failures, others are more difficult to evaluate. For instance, the myopia of the Hubble Space Telescope was immediately obvious, as was the success of the expensive corrective lens that was installed to fix it. But scientific and technical successes are not always easily demonstrated. This is one reason why so many people consume vitamin C, though there is little proof that it does any good.

If the market is working well, then, advances will proceed at as fast a rate as human ingenuity can muster. Conversely, when markets fail, science and technology will meander or stagnate. Market "failure" means, in a nutshell, that "truth has not won out." Science books are filled with examples in which truth did win out. Truth wins out as well when bad science is exposed or findings are claimed but cannot be replicated. Cold fusion is a particularly salient example of this.

The law, of course, is a major consumer of many scientific and pseudoscientific products. Sometimes it is the only consumer. Its buying habits, therefore, hugely affect what science is done and how it is done. At the same time, the law is an intensely practical discipline. As a consumer, therefore, it must sometimes settle for what is available, though it might prefer to wait for the next decade's improvements. These conflicting forces, demanding the best and settling for the best that can be done, should inform the admission of expert testimony.

Trial court gatekeepers should be especially vigilant when considering specialties whose application outside the courtroom either differs significantly from its relevance to the litigation or has no application outside the courthouse. Rape trauma syndrome, for instance, was initially proposed as a therapeutic tool, not as a means to identify rape victims. Post-traumatic stress disorder has a similar genesis. Another example is the polygraph test. These "lie detectors" have been the subject of intense research attention, though with uncertain results. While polygraphs are used in many settings, especially in the employment context, their use in the courtroom is special. Outside the courtroom, the polygraph has utility even if its results are not particularly valid. The threat of the polygraph alone might be sufficient to achieve such goals as deterring theft or compelling a confession. As then President Richard Nixon observed, caught on one of the infamous Watergate tapes, "Polygraphs? I don't know if the damn things work, but I do know that they scare the hell out of anyone who has to take one." In the courtroom, however, polygraph, results are typically offered as substantive evidence. When a defendant "fails" a polygraph, a reasonable juror is likely to conclude that the defendant "did it." If the polygraph is not a valid device, however, it should not be allowed. As the main consumer of this technology, judges should insist that it live up to its promises.

Perhaps the best modern example of a technique specially developed for the courtroom but based on a more general technology is DNA profiling, sometimes also misleadingly referred to as DNA fingerprinting. This technology is a model for innovations in forensic science. DNA profiling was built on scientific fundamentals of microbiology and population genetics that were expanded for the law's use. At the start, both the legal and the lay communities responded with great enthusiasm to this technology. It was a miracle means for unambiguously identifying the culprit in many kinds of crimes. Everyone's DNA is unique, with the possible exception of that of identical twins. In addition, and of great practical significance, DNA can be recovered from a wide range of bodily products, such as blood, hair, sperm, fingernails, and saliva, products that perpetrators routinely leave at the scene. The crime-fighting potential for DNA profiling is vast. For example, in the investigation of the "Unabomber," the FBI managed to obtain a DNA profile matching Ted Kacynzski's from the saliva left on the stamp affixed to one of the bomber's letters. Current DNA technology means that perpetrators should refrain from leaving hair, saliva, blood, fingernails, and other bodily secretions or excretions at the scene; this is not easily accomplished.

But even a technology as sophisticated as DNA profiling is not free of the possible corruption of human influence. Unfortunately, DNA is now associated with the trial spectacle of the late twentieth century, the prosecution and acquittal of O. J. Simpson. The DNA evidence in that case, possessed in abundance by the prosecution, was insufficient to secure a conviction. A popular reaction to the large role DNA played in the prosecution was that perhaps DNA is not as impressive as once thought. The Simpson defense, however, never challenged the theory or technology of DNA profiling. In fact, Simpson's two main DNA experts, Peter Neufeldt and Barry Scheck, regularly rely on DNA technology in their work for criminal defendants. In the Simpson defense strategy, it was the human element that received all the attention.

The Simpson defense, in particular, offered two theories of what went wrong with the DNA evidence. First, they attacked the police handling of the evidence, suggesting that the police had the motive and opportunity to plant Simpson's blood at the scene. Under this theory, it should have been no surprise to the police to find that Simpson's blood matched blood found at the scene, since they had put it there. The DNA test, therefore, was 100 percent accurate; the test was not corrupt, the police were. The second theory of human error concerned the processing of the blood at the laboratory. This theory contemplated the possibilities of both fraud and incompetence. Again, there was nothing wrong with the test that a little professional integrity could not have cured.

No scientific technique, however, can guarantee against the failings of human nature. In fact, to the extent that science casts an aura of objectivity over the evidence, special solicitude is needed to guard against the potential for human error, whether purposeful or merely negligent. But the possibility that a scientific technique might be abused should not lead to either of two extreme positions. The courts should not admit all pseudoscience on the theory that all expertise is fallible and thus bad science is not different in kind from good science. Nor can courts exclude good science because people are inherently fallible. The challenge for judges and the law lies in understanding the science well enough to make policy judgments in light of the possibility of error. Although we cannot expect judges to get all answers involving scientific matters correct, we should expect at least that they understand the questions.

What Is to Be Done?

As the Supreme Court in *Kumho* noted, "There are many different kinds of experts and many different kinds of expertise."[30] The Court, however, refrained from offering any prescription regarding how these many different kinds of expertise might be evaluated by trial courts. It explained that "too much depends on the particular circumstances of the particular case at issue."[31] Thus the courts should be expected to begin articulating sets of factors that might constitute "reasonable measures of the reliability" of particular forms of expert testimony.[32] As a general matter, the various types of expert testimony courts see can be divided roughly into five categories.

In the first category are experts who propose to testify to a general or specific scientific opinion that is supported or refuted by sound research. The theory of DNA technology is one example. Other examples include the effects of hypnosis on memory, the relationship between smoking cigarettes and lung cancer, and the lack of a relationship between Bendectin and birth defects. This category is the paradigmatic example of scientific evidence as discussed in *Daubert*. When the research exists, the expert testimony is admitted. When it is not available or it is poorly done, the testimony is excluded.

The second category includes experts who propose to testify to an opinion that is based on experience but has not been studied extensively or at all. However, it could be studied in a systematic and rigorous fashion. Two examples I have considered at length fit into this category, the battered woman syndrome and handwriting identification. The category squarely presents the question whether the experience of the expert is enough. It is important, however, that courts bear in mind that their answer is likely to affect the amount and quality of research that is forthcoming. If the decision is that experience is enough, then for many fields nothing more will be done.

In the third category, experts propose to testify to an opinion based on experience that is inconsistent with other research but that has yet to be studied fully. The respective subjects, however, could be studied systematically and rigorously. This category is similar to category one but with the twist that the experience of the expert is inconsistent with general research or theoretical expectations. The theory of repressed memories that put George Franklin in prison for seven years is one example of this group. Another example is the theory that electromagnetic fields cause childhood leukemia. The *Daubert* Court's example of the effects of a full moon on human behavior would also fall into this category. Once again, the admissibility decision will often affect the professional community from which the expert comes.

Experts in the fourth category propose to testify to the application of a well-researched or well-regarded body of work but are not themselves familiar with the underlying theoretical or research basis for the work. Also in this category would be experts who testify to immediate sense experiences to which they should have received adequate and accurate feedback. Auto mechanics, electricians, plumbers, accountants, some engineers, and some DNA technicians fall into this group. The physician who testifies about the clamp that does not leak and the harbor pilot who does not ground ships on sandbars would both also fit here. These experts are paradigmatic examples of "technicians." It is not that the underlying basis of their testimony is not "science" or has not been (or could not be) tested. In fact, ordinarily, the opposite is true. But there is no need for it to be demonstrated over and over again in court. Moreover, opponents of the evidence can generally challenge the application of the technology with their own experts.

The fifth and final category includes experts who propose to testify to opinions based on experience that is not testable but that comes from a field with certain professional standards. Examples include most historians or an art expert who proposes to identify a forgery on the basis of brushwork. Experts in literature and theology would also fall in this category. These experts are paradigmatic examples of "specialists." Courts should defer, to some degree, to professional standards of competence, but they must decide for themselves which standards and which experts are acceptable. In *Daubert*, for instance, the Court relied on Karl Popper, not Paul Feyerabend, as its choice of a philosopher of science. And in *Kumho*, the Court emphasized that general acceptance cannot "help show that an expert's testimony is reliable where the discipline itself lacks reliability, as, for example, do theories grounded in any so-called generally accepted principles of astrology or necromancy."[33] Other fields present similar philosophical choices.

Crafting admissibility standards that apply consistently across professional disciplines will not be an easy task. It will also probably take considerable time and effort. In pursuing this goal, courts should keep in mind two key lessons from *Daubert*. First, in making judges gatekeepers, the Court made clear that judges should not abdicate responsibility for determining what is worthwhile testimony to expert communities. The second lesson, though only implicit in *Daubert*, was that courts should, wherever possible, adopt admissibility criteria that encourage expert communities to develop the best possible information on legally relevant issues. Courts cannot be passive consumers of whatever opinions expert communities decide to produce. Although in the short term this might play havoc with certain fields, in the end both they and the justice system will be better for it.

Into a Brave New World

Although judges will not soon be joining physicists in coming up with a grand unified theory, they should be expected to have a solid understanding of the scientific method. There are significant costs associated with both admitting bad science into the courtroom and excluding good science from the courtroom. The United States Supreme Court, recognizing this basic insight, established judges as "gatekeepers" and made them responsible for evaluating scientific merit at the threshold of their courtrooms. Although seemingly overwhelming, this responsibility has become a necessary component of a judge's job description. In deciding the admissibility of expert testimony, therefore, the validity of polygraph results, the value of epidemiological research on the effects of silicone gel, and the scientific basis for the battered woman syndrome are all matters that affect the justice of the final outcomes. Judges have always prided themselves as perhaps the last great generalists in a world of increasing specialization. Judges have always had to have a solid education in the liberal arts, whether it involved the historical basis of the First Amendment or the economic theory underlying antitrust law. As we move into the twenty-first century, this liberal arts orientation must also include math (especially statistics) and science. No one can be a true generalist today without a solid understanding of these subjects.

As judges begin to exercise their facility with distinguishing science from pseudoscience, they will find that some formerly accepted experts fail to meet even the most minimal of standards. Today, for example, Freudian psychologists are routinely permitted to testify. But an expert who bases his testimony on Dostoevsky's great works would not be permitted to testify. Yet Freud's work is certainly no more scientific than Dostoevsky's novels, and indeed Dostoevsky might have been a more percipient observer of human behavior than Freud. Any principled rule of admissibility should treat Freudian psychology the same as it treats Dostoevskian psychology. And if that rule is informed by the scientific method, both should be excluded from the witness stand.

Judges must become sophisticated consumers of science. However, as consumers, judges need not themselves be scientists. The scientific sea is wide and deep, and judges cannot be expected to be steeped in the detail of all the science that comes before them. At the same time, although judges need not be expert enough to write scientific articles, they should be proficient enough to read them. As judges begin to plunge into the scientific sea in an effort to learn enough to stay afloat, they will find that it is a rough and sometimes inhospitable environment. But with practice and patience, they should find that, despite the coldness of the waters, science contains a world of immeasurable value to the law. Indeed, science is essential for understanding the world the law is charged with regulating.

IT IS SO, IF THE SUPREME COURT THINKS SO

The Supreme Court's Use of Science in Constitutional Interpretation

A constitution, to contain an accurate detail of all the subdivisions of which its great powers will admit, and of all the means by which they may be carried into execution, would partake of the prolixity of a legal code, and could scarcely be embraced by the human mind. It would probably never be understood by the public. Its nature, therefore, requires, that only its great outlines should be marked, its important objects designated, and the minor ingredients which compose those objects be deduced from the nature of the objects themselves. . . . In considering this question, then, we must never forget that it is a constitution we are expounding.

—JOHN MARSHALL, *McCulloch v. Maryland*

Law is through and through a social phenomenon; social in origin, in purpose or end, and in application.

—JOHN DEWEY

The judicial act of interpreting the United States Constitution is shrouded in mystery. Among those who comment on such matters, the interpretive process is regarded in wildly different ways. In one view, constitutional interpretation is limited to a mechanical reading of the plain text. In another, it is an act of faith not unlike biblical interpretation. In a third, it is an act of political will in the search for the "just" society. Supreme Court justices, therefore, whose task it is to interpret the Constitution, might be described as mere technicians, high priests, or benevolent despots. These job descriptions overlap little, creating much uncertainty about the Court's role in the American political drama. Complicating matters and potentially undermining the Court's very legitimacy, is the fact that the Supreme Court's role in evaluating the constitutionality of the actions of the Congress and the president (as well as the states) is not explicitly established by the Constitution. The invention of judicial review did not come about until fourteen years after the ratification of the Constitution, and it would not be exercised again to overturn an act of Congress for another fifty-four years. In its early years, the Court had little power, real or supposed. Indeed, John Jay, the first chief justice, resigned after being elected governor of New York, exclaiming that the Court lacked "energy, weight, and dignity." Today, of course, the more likely transition is from governor to chief justice.

The modern respect, even reverence, we hold for the Supreme Court is largely attributable to the fourth chief justice, John Marshall, and the invention of judicial review. Although Marshall did not invent the practice of judicial review, he engineered the coup that set it at the core of constitutional jurisprudence. It is a story often told and worth repeating.

George Washington had hoped that the fledgling nation would be free from the strife he considered the inevitable product of political parties. This proved unrealistic, for although the country had largely united behind Washington, it soon divided between the Federalists of John Adams and what would become the Republican party of Thomas Jefferson. After Washington's self-imposed retirement following his second term in office, John Adams was elected president. Jefferson, second in the polling, became vice president. John Marshall served as secretary of state in the Adams administration. Although serving the same administration and related by blood—they were second cousins—Jefferson and Marshall were political enemies. Henry Adams described the feelings that existed between these two men in this portrait of Marshall: "This great man nourished one weakness. Pure in life; broad in mind, and the despair of bench and bar for the unswerving certainty of his legal method; almost idolized by those who stood nearest him . . . this excellent and amiable man clung to one rooted prejudice: he detested Thomas Jefferson. . . . No argument or entreaty affected his conviction that Jefferson was not an honest man."[1] Jefferson felt similarly about Marshall. Their animosity toward one another would build to a crescendo shortly after the election of 1800, the year Jefferson was elected president.

In reality, the transfer of power between Adams's Federalists and Jefferson's Republicans was the first genuine transfer of power between rival factions under the still fledgling Constitution. The only prior transfer of power, from Washington to Adams, was certainly historic, but Adams was ideologically aligned with Washington. In contrast, Jefferson's ascension to power was revolutionary, because of the existing political rivalry. Between the moment of Jefferson's election in November 1800 and the time he took the oath of office in March 1801, Adams did all he could to entrench Federalists in power. Adams began by nominating Marshall to be the fourth chief justice. Marshall, however, did not immediately give up his duties as secretary of state. Before leaving office, Adams and the Federalist Congress substantially expanded the size of the federal judiciary, and Adams labored strenuously to fill the new posts with Federalist judges before Jefferson took power. Marshall, as secretary of State, was responsible for signing and delivering the commissions to the new judges. These "midnight judges" were the subject of intense criticism by the Jeffersonians. In the rush of business in March 1801, some of the commissions remained undelivered before Adams and Marshall left office. William Marbury was one of the unfortunates who failed to receive his commission.[2]

Jefferson and his newly appointed secretary of state, James Madison, claimed that the undelivered commissions were nullities. Marbury brought suit directly in the Supreme Court asking the Court to order Madison to deliver the commission. In *Marbury* v. *Madison*,[3] John Marshall, writing for the Court, held that Marbury indeed had a right to his commission and that the secretary of State had an obligation to deliver it to him. However, this is not the enduring significance of *Marbury*. In a brilliant rhetorical sleight of hand, Marshall explained that the Supreme Court lacked the legislative authorization to issue the order Marbury sought because the congressional act purporting to give the Court this power violated Article III of the Constitution. Thus, though ostensibly holding for Madison, the Court claimed the constitutional authority to review the actions of both the president and the Congress, thereby establishing the institution of judicial review. As Marshall interpreted the Constitution, the Court was obliged to ensure that the president acted within constitutional bounds and, as determined here,

> should Congress, under the pretext of executing its power, pass laws for the accomplishment of objects not entrusted to the government; it would become the painful duty of this tribunal . . . to say that such an act was not the law of the land.[4]

Because the Constitution was silent on the power of judicial review, Marshall needed to produce a variety of justifications for finding this power. Marshall argued in *Marbury* that principles implicit in the text, historical precedent, and the framers' decision to propound a written Constitution all supported judicial review. In addition, and central to his overriding argument, Marshall made the factual assertion that, in general, legislators are more inclined than the judiciary to stray from the Constitution. Relying on this factual assertion, Marshall explained that the Constitution forms "the fundamental and paramount law of the Nation."[5] It follows that "it is a proposition too plain to be contested, that the constitution controls any legislative act repugnant to it."[6] It was equally plain that the judiciary must be the ultimate arbiter of what the Constitution says, for without judicial review the legislature could "alter the constitution by an ordinary act."[7] The Constitution, therefore, established (albeit not expressly) the Court as the final arbiter of constitutional meaning because institutionally it was less likely to abuse that power than the Congress or president.

Jefferson was aghast at Marshall's pronouncement that the Court wields the power to review the constitutionality of the acts of the coordinate branches of government. Jefferson considered judicial review to be inconsistent with a democratic republic and a threat to American liberties. Jefferson detailed his objections to the institution of judicial review in a letter to William Jarvis:

> You seem . . . to consider the judges as the ultimate arbiters of all constitutional questions; a very dangerous doctrine indeed, and one which would place us under the despotism of an oligarchy. Our judges are as honest as other men, and not more so. They have, with others, the same passions for party, for power, and the privilege of their corps. Their maxim is *"boni judicis est ampliare jurisdictionem,"* and their power the more dangerous as they are in office for life, and not responsible, as the other functionaries are, to the elective control. The constitution has erected no such single tribunal, knowing that to whatever hands confided, with the corruptions of time and party, its members would become despots. It has more wisely made all the departments co-equal and co-sovereign within themselves.[8]

As has been the case in many other debates, history has failed to side with Jefferson here. Although the Court struck down only one congressional act between *Marbury* and the Civil War—and that one, *Dred Scott*, contributed to the war's onset—the twentieth century has seen a Court that regularly exercises that power.

Divining the Meaning of the Constitution

The Court as the ultimate determiner of the meaning of the Constitution leads to the question of just how the Court can know what the Constitution means. Constitutional law is beset by a monumental difficulty not confronted by most other legal subjects. Whereas contract law is largely about the Uniform Commercial Code and criminal law is largely about state or federal criminal codes, few believe that constitutional law has very much to do with the Constitution. Constitutional interpretation is only barely associated with the words of the document that resides in the National Archives. The Constitution is primarily about governing. Constitutional adjudication is about taking the mostly ambiguous words of the Constitution and applying them to concrete cases that were only dimly, if at all, contemplated by the drafters of the document. In this task, the justices need significant assistance from outside authorities.

The Supreme Court historically has relied on a wide variety of resources, or authorities, to give meaning to the hopelessly ambiguous text. The terms, for example, *equal protection* and *due process* are not self-defining. Even the meaning and scope of a term as seemingly straightforward as *free speech* is not plain. Does it include child pornography and commercial speech and, if so, to the same extent as political speech? More than the text alone is needed to give depth to the words. In fact, no judge or scholar doubts that authorities beyond the text must be consulted to give meaning to the Constitution. The Court traditionally relies on four sources of authority beyond the text to divine the principles of constitutional law: the original intentions of the drafters of the Constitution, prior case law (also known as precedent), constitutional scholarship, and contemporary values. Another often ignored source of authority that contributes to the Court's interpretive process is the fact-rich world in which the principles and values of the Constitution must operate. Many of these facts are or could be the product of scientific research. These five sources, however, do not carry equal weight, nor are they equally helpful in providing guidance on the Constitution's meaning.

Judges and scholars uniformly agree that the original intent of the drafters can shed needed light on the meaning behind the Constitution's words. Courts often operationalize original intent by considering the drafters' intentions or, more ambitiously, the intentions of those who voted to ratify it. But subjective intentions are difficult to discern in individuals, let alone groups of legislators or the voting populace. Consequently, this standard provides little concrete assistance. More important, a Constitution intended to endure forever must be interpreted to respond to contingencies and changing circumstances far beyond those imagined by even the most far-sighted of its drafters. A series of cases decided shortly after the Civil War illustrates this point well. In those cases, the Court was asked to determine the constitutionality of Congress's decision to print paper money. As for the drafters' original intent there was no question: Congress had only the power "to coin money." Luther Martin, a delegate to the Constitutional Convention, explained in 1787: "A majority of the convention, being wise beyond every event, and being willing to risk any political evil, rather than admit the idea of a paper emission, in any possible event, refused to trust this authority to [the federal] government."[9] By 1871, however, when the *Legal Tender Cases* came before the Court, the power to print paper money, a power held by all sovereign nations, had become indispensable. The Court, ignoring both the plain intent of the drafters and a one-year-old precedent, held that Congress's power extended to the printing of "legal tender for the payment of all debts."[10] As many other cases like the *Legal Tender Cases* demonstrate, although original intent often provides useful insights, it rarely provides ready answers to the meaning of constitutional provisions, and the Court ignores it with some alacrity.

Another source of generally accepted authority is the Supreme Court's own precedent (or the principle of *stare decisis*). But, as the *Legal Tender Cases* illustrate, prior decisions do not bind the Court's hands too severely. In many instances, precedent is ambiguous or not clearly pertinent to new circumstances. More important, because the Supreme Court is the only institution (short of constitutional amendment) that can overturn its own precedent, blind adherence to the principle of *stare decisis* would straitjacket constitutional jurisprudence. It would make the justices of today mere enforcers of the will of justices now long dead. Justice Felix Frankfurter explained the Court's usual view of precedent:

> We recognize that *stare decisis* embodies an important social policy: It represents an element of continuity in law, and is rooted in the psychologic need to satisfy reasonable expectations. But *stare decisis* is a principle of policy and not a mechanical formula of adherence to the latest decision, however recent and questionable, when such adherence involves collision with a prior doctrine more embracing in its scope, intrinsically sounder, and verified by experience. . . . This Court, unlike the House of Lords, has from the beginning rejected a doctrine of disability at self-correction.[11]

Precedent typically provides a good starting point for constitutional analysis. Yet the Court rarely follows precedent fully, and never uncritically.

Still another source of authority is constitutional scholarship, usually furnished by law professors. Indeed, throughout history many justices were themselves professors before joining the Court, including three now on the Court: Justices Scalia, Kennedy, and Ginsberg. Constitutional scholarship includes both examinations of conventional views of the structure and meaning of constitutional government and excursions into broader principles of philosophy, including exegeses on theories of justice. The Supreme Court, though not always explicit about its reliance on constitutional scholarship, often supports and buttresses its opinions with the ruminations of law professors. Examples of particularly influential constitutional scholars in the latter half of this century include Herbert Wechsler, Gerald Gunther, Laurence Tribe, and John Hart Ely. The very busy justices are thus aided by the reflections of the more leisurely professorial class.

The final conventional source of authority for constitutional adjudication, and the most controversial, is the moral sense of contemporary society. Contemporary values are sometimes used explicitly, as in the context of the Eighth Amendment, where the Court contemplates contemporary practices in determining whether, for instance, the death penalty is cruel and unusual punishment. Just recently, a preeminent group of moral and political philosophers—Ronald Dworkin, Thomas Nagel, Robert Nozick, John Rawls, Thomas Scanlon, and Judith Jarvis Thompson—submitted an *amicus curiae* ("friend of the Court") brief in the two cases discussed in Chapter III that raised the issue of whether individuals have a constitutional right to "physician-assisted suicide." The brief argued, among other things, that the principle of individual autonomy, a principle the Court has deemed a part of the liberty protected by due process, "encompasses the right to exercise some control over the time and manner of one's death." The brief dismissed the suggestion that, as moral philosophers, they were inviting "the Court to make moral, ethical or religious judgments." However, this conclusion is inescapable. Indeed, how could the Court reconcile or even recognize a state's legitimate interest in protecting life and the individual's right to control his own death without rendering some moral judgment? Although it remains fashionable for the Court to denounce the relevance of contemporary values in constitutional adjudication, such values play a central role in fixing the scope and content of the Constitution's boundaries.

While judges and scholars have not typically identified facts as a source of *authority* for determining constitutional meaning, facts play, and historically have played, a pivotal role in exactly this capacity. This is not to say that observers have failed to notice the pervasive nature of fact-based inquiries in constitutional decisions. Indeed, no half-alert Constitution watcher could miss their brooding omnipresence. But judges and lawyers have mostly failed to appreciate their distinctive role in constitutional adjudication. Throughout its history, factual assertions have supplied an alternative basis for Court decisions in much the same way original intent or precedent might. Facts thus serve the interpretive function of the Court. The justices use, and have always used factual assertions as rhetorical flourishes that are measured by their power to persuade. Consequently, the Court has required little independent validation of facts as might have been or failed to have been substantiated by rigorous research.

Legal observers have yet to fully appreciate the power of facts in this interpretive context. More troubling, the Court itself appears to little appreciate the significance of fact finding and the scientific method for constitutional jurisprudence. Thus, empirical research has served as a nominal source of authority for constitutional interpretation, but it has not been a source used with any consistency or sophistication by the Court. A proper understanding of science would lead to a profoundly different role for facts in the enterprise of divining constitutional meaning. This is the story of a constitutional world enmeshed in the darkness of ignorance of science. In holding a light on this darkness, I hope to demonstrate why the Court must escape its ignorance and, in the process, possibly illuminate the path the Court should take in making its escape.

"Give me liberty or . . . ": The Problem of "Substantive Due Process"

> Is life so dear or peace so sweet, as to be purchased at the price of chains and slavery? Forbid it, Almighty God! I know not what course others may take, but as for me, give me liberty or give me death!
>
> —PATRICK HENRY (1775)

To this day, the lasting legacy of *Marbury* v. *Madison* is the concern that the Court will too readily wield the power of judicial review to nullify the will of the majority. As Judge Learned Hand stated, "For myself it would be most irksome to be ruled by a bevy of Platonic Guardians, even if I knew how to choose them, which I assuredly do not."[12] At the same time, we have come to accept the Court as the guarantor of liberty, meaning the right of an individual, sometimes, to "be let alone." As our history too well attests, from the slander laws of the eighteenth century to the bans on flag burning of more recent times, the majority sometimes demonstrates little tolerance for the excesses occasionally associated with liberty.

Much of constitutional law involves the Court in navigating this dilemma of reconciling majority will and individual liberty.

When the fear is judicial activism, the bogeyman of constitutional law is *Lochner* v. *New York*.[13] In *Lochner*, the Court was asked to interpret the due process clause of the Fourteenth Amendment to determine whether it prohibited the New York legislature from passing a law restricting the number of hours that bakers could work. Ultimately, the Court found that the New York law violated the parties' liberty of contract. Justice Peckham, writing for the Court, concluded, "It is manifest to us that the [New York law] . . . has no such direct relation to and no such substantial effect upon the health of the employe [sic], as to justify us in regarding [it] as really a health law."[14] The Court thus rejected the New York legislature's judgment and factual determination that shorter work hours for bakers were necessary to their and society's health. Instead, Peckham found the act to be "an illegal interference with the rights of individuals, both employers and employes [sic], to make contracts regarding labor upon such terms as they may think best."[15] Two factual assumptions thus underlie the *Lochner* decision. First, the Court rejected the state's factual assertion that shorter hours would lead to better health. Second, the Court believed in the existence of equal bargaining power between employer and employee.

Lochner is usefully contrasted with *Muller* v. *Oregon*[16] decided just three years later. In *Muller*, the Court was asked to interpret the same constitutional provision at issue in *Lochner* to determine whether it was violated by an Oregon statute limiting the work hours for women employed "in any mechanical establishment, or factory, or laundry."[17] The Court reached the opposite conclusion and upheld the Oregon law. The *Muller* Court found that the Oregon statute was reasonable and would contribute to the health of the women workers. Significantly, supporting the opinion that women's health required shorter work hours was a brief submitted by Louis Brandeis, who was later appointed to the Supreme Court. This brief, thereafter used to define a class of briefs known as "Brandeis briefs," provided an exhaustive summary of the social science of the day that demonstrated, in the words of the brief, that the "overwork of future mothers . . . directly attacks the welfare of the nation." Contrary to New York in *Lochner*, then, Oregon apparently persuaded the Court in *Muller* that its maximum hour law would promote good health.

But the two cases cannot be reconciled simply on the basis that Oregon had better social scientists or lawyers than did New York. Indeed, as Justice Harlan's dissenting opinion in *Lochner* indicates, the Court had a multitude of social science authorities before it that demonstrated that the long and difficult hours of New York bakers adversely affected their health. Instead, *Lochner* and *Muller* can be reconciled because they share a common theory of human nature. This theory, generally associated with Social Darwinism as popularized in the works of Herbert Spencer, dictated that freedom of contract was to be protected against legislative intrusion except in certain categories of cases. The men in New York had to fend for themselves, while the women in Oregon received the Court's beneficence. The Court in *Lochner and Muller* thus considered state-sponsored social engineering justified only when the state sought to protect "the weaker sex." Justice Oliver Wendell Holmes, in his *Lochner* dissent, captured the driving premise of the Court's conclusion in his characteristically potent fashion:

> This case is decided upon an economic theory which a large part of the country does not entertain. . . . The Fourteenth Amendment does not enact Mr. Herbert Spencer's Social Statics.[18]

At one level, Justice Holmes's *Lochner* dissent can be interpreted as imploring the Court to set aside its personal sociology in order to properly interpret the Constitution. At another level, however, Holmes's dissent can be challenged as a product of *his* particular beliefs about human relations. Holmes, as perhaps the first legal realist, openly embraced a fact-based jurisprudence as indicated by his often-quoted statement: "The life of the law has not been logic: it has been experience."[19] That Holmes discounted the relevance of Herbert Spencer in deciding *Lochner* belies the importance of extraconstitutional considerations to his conclusion. The social milieu continually shaped Holmes's constitutional outlook. Holmes himself believed that all "rules of law presuppose a certain state of facts to which they are applicable."[20] In fact, it has been argued that Holmes's *Lochner* dissent was shaped by his own particular brand of Social Darwinism. Holmes openly embraced a political philosophy that accepted the principle we today would term "might makes right": "If the will of the majority is unmistakable, and the majority is strong enough to have a clear power to enforce its will, and intends to do so, the courts must yield . . . because the foundation of sovereignty is power, real or supposed."[21] Holmes's judicial deference to legislatures depended as much on his sociological and psychological assumptions as on any rationale he found in the text of the Constitution.

Holmes's most infamous embrace of a fact-based jurisprudence comes from his opinion in *Buck v. Bell*.[22] In *Buck* the Court was asked to determine whether a Virginia statute mandating in certain cases "the sterilization of mental defectives" violated the due process clause.[23] The

Court held that it did not. Although this holding coincides with Holmes's long-standing philosophy of deferring to legislatures, his opinion for the Court clearly reflects his acceptance of the scientific theory of eugenics on which the legislation was based. Holmes's own chilling words make the point:

> We have seen more than once that the public welfare may call upon the best citizens for their lives. It would be strange if it could not call upon those who already sap the strength of the State for these lesser sacrifices, often not felt to be such by those concerned. It is better for all the world, if instead of waiting to execute degenerate offspring for crime, or to let them starve for their imbecility, society can prevent those who are manifestly unfit from continuing their kind. The principle that sustains compulsory vaccination is broad enough to cover cutting the Fallopian tubes. Three generations of imbeciles are enough.[24]

Modern commentators' reactions to cases such as *Lochner* and *Buck* are somewhat complicated. Virtually all agree that the factual assumptions underlying these decisions are wrong. However, the condemnation of these decisions appears to require more and typically engenders more. The not surprising fear is that while we know these facts are false, might they not be accepted again at some future time? This led some commentators to condemn the whole enterprise of constitutional fact review and the constitutional rules that make such review possible or necessary. But this has proved impossible. Since legal principles transcend cases, they must be preserved for the cases the critics hold dear. Hence, for example, a conservative Court might wish to retain power to review factual findings that support liberal legislative enactments, whereas a liberal Court would want to employ the same rule to review conservative rule making. Eliminating the rules that engender constitutional fact review, then, can boomerang if they are not available when you need them later. Manipulating the facts is so much more convenient. But this tactic only succeeds in consistently producing the desired results as long as the facts can be "found" with some freedom from empirical reality. As the sciences improve, they increasingly remove this freedom from courts.

"We hold these truths to be self-evident . . . "

Although facts have become more important over time, the Court has demonstrated little improvement in its employment of empirical research in deciding constitutional cases. In 1803, it took four days by carriage to travel from Boston to Washington, D.C., while in 1973 it took about the same amount of time to travel from the earth to the moon aboard an Apollo spacecraft. Yet the Court on which Thurgood Marshall sat in 1973 had little better understanding of the scientific method than the 1803 Court of John Marshall. And there has been no improvement since 1973.

" . . . that all men are created equal . . . "

Constitutional scholars typically fix "the modern period of American constitutional law to be the period since *Brown* v. *Board of Education*."[25] Conveniently, *Brown*[26] also marks the modern era of the Court's explicit use of scientific research in constitutional law.[27] In *Brown*, the Court was asked to interpret the scope of the Fourteenth Amendment's guarantee of "equal protection" in the context of racial segregation in public schools. In its opinion, the Court cited a series of studies conducted by Dr. Kenneth Clark and others to support its finding that segregation of blacks "generates a feeling of inferiority as to their status in the community that may affect their hearts and minds in a way unlikely ever to be undone."[28] Consequently, the Court concluded that segregation and its principle of "separate but equal" violated the equality guaranteed by the Fourteenth Amendment. This apparent reliance on social science was applauded by some as marking the inception of the modern era of cooperation between social science and law[29] and criticized by others who feared the consequences of relying too heavily on science.[30] Still others have questioned how much influence the research had on the Court's decision.[31] In retrospect, it seems clear that the studies were not necessary to the holding, and, indeed, at the time, Chief Justice Earl Warren is reputed to have responded to the controversy surrounding the citation to the studies by saying, "It was only a note, after all."[32]

It might reasonably be asked, then, why did the *Brown* Court rely on social science at all, rather than the "bedrock of a coherent constitutional principle"?[33] The simple answer might be that the Court interpreted the clause as raising the question whether "segregation of children in public schools solely on the basis of race . . . deprive[s] the children of the minority group of equal educational opportunities."[34] Relevant to this inquiry was the "feeling of inferiority"[35] segregation instilled and the detrimental effects engendered by segregation. These are empirical questions, and however obvious the answers might seem, social science was pertinent to the *Brown* inquiry. Still, this "simple explanation" does not reveal why the Court interpreted the equal protection clause as raising an empirical question in the first place.

The answer to the question "Why social science?" comes from considering the question "What else was available?" Commentators have offered a variety of alternative bases for deciding *Brown* so that the result comes out the same, but few agree on any single one. Indeed, constitutional theorists uniformly deem it necessary to provide a plausible account of *Brown* when explaining their overall constitutional theory. However, if one looks to the four traditional sources of authority relied on for constitutional adjudication, it becomes clear that not one squarely supports the decision and several indicate a contrary result.

The text of the Equal Protection Clause of the Fourteenth Amendment is, at best, ambiguous on the matter, and the *Brown* result appears contrary to the framers' original intent. Segregation was widely practiced after adoption of the Fourteenth Amendment, and, in fact, it was the practice in the District of Columbia itself when the amendment was drafted. Although precedent shows a slow movement toward the *Brown* result, the doctrine of "separate but equal" was firmly established in the prior case law of *Plessy* v. *Ferguson.* Moreover, at the time, constitutional scholarship was divided about the proper outcome, and contemporary values in 1954 did not support the decision. Given the shortcomings of the traditional sources of authority, the Court not surprisingly embraced constitutional fact-finding to support its ruling.

That the factual insight was indispensable to the immediate holding in *Brown* is plain. Still, serious doubt attaches to whether the Court truly relied on the social science research for this insight. The simplest way to determine just how seriously the Court considered the factual showing is to ask whether the result would have been different if the evidence had shown the contrary. Suppose empirical research demonstrated that the quality of education in segregated schools was substantially *superior* to an integrated education. This is exactly the assertion made by all-black colleges and women-only schools today. We need not speculate about this issue, since this very claim was adjudicated about ten years after *Brown.*

In *Stell* v. *Savannah-Chatham County Board of Education,*[36] the plaintiffs brought suit challenging the continued operation of the defendant's "biracial system." The defendants produced experts, including Dr. Ernest van den Haag, Professor of Social Philosophy at New York University, willing to testify that research indicates that integrated education leads to increased tension and decreased performance rates. The plaintiffs complained that this research was irrelevant, since "the law is settled by the Supreme Court in the *Brown* case that segregation itself injures Negro children in the school system." Thus, the plaintiffs asserted, they "do not have to prove that." The trial court disagreed. The court found that the factual judgment of *Brown* was integral to the decision but might be different in other communities or in other historical times. The court explained its understanding of *Brown*'s finding that segregation deleteriously affects black self-esteem and interferes with their educational and mental development:

> These are facts, not law. . . . Whether Negroes in Kansas believed that separate schooling denoted inferiority, whether a sense of inferiority affected their motivation to learn and whether motivation to learn was increased or diminished by segregation was a question requiring evidence for decision. That was as much a subject for scientific inquiry as the braking distance required to stop a two-ton truck moving at ten miles an hour on dry concrete.[37]

The plaintiffs appealed to the United States Court of Appeals for the Fifth Circuit.[38] In an opinion by Judge Griffin Bell (later appointed attorney general under President Jimmy Carter), the appellate court made short work of reversing the trial court's decision. Judge Bell stated that *Brown* was not "limited to the facts of the cases there presented" and that the major premise of *Brown* was a constitutional principle, not a factual determination: "We read [*Brown*] as proscribing segregation in the public education process on the stated ground that separate but equal schools for the races were inherently unequal." The Supreme Court never reviewed the decision in *Stell.* However, the fact that the Court shared Judge Bell's view of the importance of the factual premise for the *Brown* conclusion that separate but equal was unconstitutional was made clear in several decisions that followed. In a series of *per curiam* opinions after *Brown,* the Court extended its holding to public beaches, golf courses, and other public facilities without the benefit or any mention of research indicating that segregation of these facilities had detrimental effects.

The matter of the factual hypothesis in *Brown,* that segregation injures black school children, is something of a constitutional muddle. At least as scientists understand facts, they are not subject to the sort of judicial fiat asserted by Judge Bell. Yet the courts seem to have a rather different conception of the empirical world. Perhaps facts are a little less factual for lawyers than they are for scientists. Professor Ronald Dworkin, a preeminent legal theorist, seems to suggest that this might indeed be so. Dworkin argues that the empirical studies cited in *Brown* were irrelevant to the question of the effects of segregation. Beginning with a premise from another scholar, Dworkin explains:

> "We don't need evidence for the proposition that segregation is an insult to the Black community—we *know* it; we know it the way we know that a cold causes snuffles." It is not that we don't need to know it nor that there isn't

something there to know. There is a fact of the matter, namely that segregation is an insult, but we need no evidence for that fact—we just know it. *It's an interpretive fact.*[39]

Dworkin's use of the term "interpretive fact" appears to encompass two separate arguments. First, Dworkin believed that the social sciences remain insufficiently valid to support constitutional rulings. Using physics as his paradigm of "real" science, he argued: "While in physics it is now thought to be an unsound judgment that rests merely on correlation between observable events unsupported by some notion of the mechanics that translate the cause to the effect, social science is only able to provide correlations without the mechanics."[40] But this is simply—even simplistically—wrong. Neither is all of physics experimental (and thus noncorrelational) nor is all of social science correlational (and thus nonexperimental). Further, correlational research is not intrinsically inferior or nonscientific compared to experimental work, especially when that work is to be applied to real-world settings. Dworkin's statement reflects a basic, indeed shocking, lack of understanding of the scientific method. It might be that physicists are, on average, more rigorous than social scientists or that the social science research relied on in *Brown* was not very good. But these arguments say nothing about whether social science can be a rigorous science and whether it sometimes is.

The second and more fundamental premise of Dworkin's "interpretive fact" concept is his apparent belief that even if valid social science were available, it should never be used to support a constitutional interpretation. Dworkin posited a legal theory he calls "creative" or "constructive" interpretation, analogizing the legal-interpretive process to writing the latest chapter of a chain novel.[41] The interpreter fits his or her interpretations into the prior chapters and at the same time extends the overall work in the "best possible" direction. The theory contemplates first that the interpreter identify the "fit" between the interpretive history and the practice being interpreted and second that the interpreter impose a "purpose on an object or practice in order to make of it the best possible example of the form or genre to which it is taken to belong."[42] The enlightened constitutional jurist, therefore, takes the constitutional story as written so far and, with the case before him or her, extends it forward in the best possible direction. But does science have a role in this process? It seems clear that scientific data can be used to answer an empirical question whose relevance depends on its "fit" with the practice being interpreted and that comports with a legal conclusion that "make[s] of it the best possible example of the form or genre to which it is taken to belong."

In *Brown* the issue of segregation's effects had been an integral component of the preceding interpretive tradition. In *Plessy*,[43] for instance, the Court assumed that the "separate but equal" doctrine provided blacks with the full and equal enjoyment of public facilities[44] and, further, that any feelings of inferiority resulted "solely because the colored race chooses to put that construction upon it."[45] Hence, the research in Brown "fit" into the interpretive tradition of considering the effects of segregation. Moreover, extending this tradition in the "best possible" direction should involve taking into account the best available research illuminating the pertinent facts. As Dworkin admitted, there is a fact of the matter. Whether that fact is supported by quality research or has interpretive significance is a separate issue.

Some of the confusion here arises out of Dworkin's relaxed use of the concept of "known facts." We may *know*, as Dworkin argued we should know "interpretively," "that a cold causes snuffles." But surely, if the "interpretive" judgment is accurate, valid scientific studies should corroborate the judgment. Science is not irrelevant for demonstrating what everyone believes to be the case, though courts might wish to relegate studies corroborating the relationship between colds and snuffles to a footnote. And, to our surprise, it might turn out that some malady only associated with colds causes snuffles, so we were wrong the whole time. Certainly researchers should not be discouraged from looking into the question on the basis that we know it to be true because we know it to be true.

One suspects that more is going on here than meets the eye. Dworkin offered what he asserted was a principled analysis of what is, in actuality, a politically expedient conclusion. The reality is that "we know" that segregation is bad because "we" refuse to permit any other conclusion. Dworkin's argument is not unlike the church's reaction to Galileo discussed in Chapter I. That segregation might be beneficial or that the earth might revolve around the sun are simply not facts that are acceptable in Dworkin's or the church's respective universes. The negative effects of segregation and the geocentric model of the universe were "interpretive facts" supporting faiths that were not going to be abandoned with the demonstration of inconvenient facts. The lesson for Dworkin, however, is the same as that for the church: science has a way of undermining the most ardently held beliefs and toppling institutions built wholly on faith.

In *Brown*, the Court willingly relied on research but with little concern for its validity. As we have seen, these facts served the redemptive need to establish the precept, but their truth was hardly necessary for redemption or our belief that *Brown* was correctly decided. Faith is sometimes strong enough to overcome inconvenient facts. In *Roe v. Wade*,[46] the case second only to Brown in defining the modern period of constitutional law, where faith is less strong, inconvenient facts have wreaked havoc on the doctrine.

" . . . that they are endowed . . . with certain unalienable rights . . . "

In *Roe* v. *Wade*, Justice Blackmun, writing for the Court, constructed the constitutional framework for a woman's fundamental right to choose an abortion free from state interference. The Court made its decision "in the light of present medical knowledge."[47] Blackmun asserted that the state's "compelling" interest in the health of the mother begins at the end of the first trimester. He explained: "This is so because of the now-established medical fact [that] until the end of the first trimester mortality in abortion may be less than mortality in normal childbirth."[48] Blackmun ruled further that "with respect to the State's important and legitimate interest in potential life, the 'compelling' point is at viability."[49] He explained: "This is so because the fetus then presumably has the capability of meaningful life outside the mother's womb."[50] Medical science, therefore, delineated the two junctures during a pregnancy—at the time mortality in abortion is no longer less than mortality at childbirth and at viability—when the state's compelling interests arose. The *Roe* Court's reliance on modern science was met with vehement criticism, wholly separate from the value choices driving the majority's opinion.

Critics complain that attaching constitutional meaning to scientific opinion, even when scientists agree, condemns the Constitution to fluctuations in meaning as scientific knowledge changes. The principal proponent of this criticism has been Justice Sandra Day O'Connor. In *Akron* v. *Akron Center for Reproductive Health*,[51] Justice O'Connor warned in dissent that, due to recent advances in medicine, linking the constitutional framework in *Roe* to medical technology has set it "on a collision course with itself."[52] By the time of *Akron*, ten years after *Roe*, abortion had become safer than childbirth through approximately the sixteenth week, and viability itself was becoming progressively earlier with advancing technology. The *Akron* Court, however, disregarded Justice O'Connor's warnings and refused to abandon the trimester scheme even as science demonstrated its continuing inapplicability. The Court insisted that the "trimester standard . . . continues to provide a reasonable *legal framework* for limiting a State's authority to regulate abortions."[53] With one stroke of the pen, the Court turned the trimester framework, a framework whose original cogency rested on a scientific basis, into a rule having a basis solely in precedent—precedent without any principle.

The fault here does not lie with the decision to build a constitutional ruling on scientific facts. The fault here lies with the Court's failure to articulate the specific constitutional relevance of the facts it chose. Blackmun never originally explained *why* the Constitution mandates the use of the scientific standard of fetus viability. Blackmun affixed the state's compelling interest in maternal health at the point in time when abortion was safer than childbirth. In 1973 this time was conveniently at the end of the first trimester. If this point in time was constitutionally mandated, then as abortions become safer, the state's interest should be postponed accordingly. Similarly, if viability is the constitutionally fixed instant when the state's interest in potential life comes into being, then as viability occurs earlier, the state's interest should advance accordingly. The constitutional significance of viability still requires explanation.

The Court, however, has so far failed to identify which constitutional values are actually implicated by the scientifically calibrated timetable identified in the abortion cases. Regarding the constitutional relevance of viability, for example, Blackmun stated: "With respect to the State's important and legitimate interest in potential life, the 'compelling' point is at viability. This is so because the fetus then presumably has the capability of meaningful life outside the mother's womb." As constitutional scholar John Hart Ely pointed out, this argument mistakes "a definition for a syllogism."[54] The juncture of viability does not have obvious constitutional significance. Certainly viability could be one of a variety of interests that the Court might find necessary to any constitutional calculation establishing the woman's rights against the state's interests. But at least absent some explanation why, viability is not the only interest. Using the trimester framework, the Court was able to finesse the perhaps intractable problem of expounding the many conflicting principles and value choices that make up the fundamental right of privacy in the abortion context. Medical science simply served as a convenient proxy for this task, at least for awhile.

In Constitution-years, *Roe*'s trimester framework survived only briefly. As this discussion suggests, *Roe* was embattled from the start and survived only barely through the 1980s, largely due to Justice O'Connor's reluctance to abandon it. Finally, nearly twenty years later, the Court repudiated the trimester framework of *Roe* in *Planned Parenthood of Southeastern Pennsylvania v. Casey.*[55] Although the trimester framework was rejected, a majority of the Court upheld "the essential holding of *Roe*." The part of *Roe* the Court found essential included a continuing role for the viability juncture. Specifically, the Court recognized "the right of the woman to choose to have an abortion before viability and to obtain it without undue interference from the State." Still, the Court did not premise its retention of the viability juncture on any statement of its constitutional significance. Instead, the Court observed that "an entire generation has come of age free to assume *Roe's* concept of liberty in defining the capacity of women to act in society, and to make reproductive decisions."[56] Hence, the Court used the public's reliance on *Roe* as precedent to support continued use of a standard that contained no principle. Significantly and typically, the Court cited no studies or other authority to support its observation that an entire generation had come of age relying on *Roe* or had otherwise made lifestyle decisions that would be disrupted by overruling *Roe*. It appears that, for the Court, this is another "interpretive fact" they "just know" to be true.

Ignorance Is No Excuse

As his opinion in *Roe* illustrates and his authorship of *Daubert* v. *Merrell Dow Pharmaceuticals Inc.* exemplifies, Justice Blackmun has played a key role in the law's use of science.[57] In fact, in 1992, Justice Blackmun was presented with a distinguished lifetime achievement award by the American Psychological Association for his use of science in his constitutional decision making. In presenting the award, among other examples cited, the organization highlighted the case of *Ballew* v. *Georgia*,[58] which contains probably the most extensive use of empirical research to be found in a Supreme Court opinion.

The Court in *Ballew* considered the matter of the constitutional relevance of jury size, a subject it had first visited eight years before in *Williams* v. *Florida*.[59] In *Williams*, the Court was asked to determine whether the Constitution requires twelve-person juries or might permit some lower number. Williams had been convicted of robbery by a jury of six.

Although the Constitution mandates the use of juries in both criminal and civil cases, it does not specify the number of people that are needed to constitute a jury. In *Williams*, the Court searched for historical precedent that might support the use of twelve, but it could find none. The Court concluded from the silent text and ambiguous history that the number twelve appeared to be born out of superstition rather than principle. It had relevance, the Court asserted, only to "mystics." Instead, the size of a jury has importance only as far as it is relevant to the jury's function. The Court explained that

> the essential feature of a jury obviously lies in the interposition between the accused and his accuser of the commonsense judgment of a group of laymen, and in the community participation and shared responsibility that results from that group's determination of guilt or innocence. The performance of this role

is not a function of the particular number of the body that makes up the jury. *To be sure, the number should probably be large enough to promote group deliberation . . . and to provide a fair possibility for obtaining a representative cross-section of the community.*[60]

Based on this analysis in *Williams*, the Court determined that conviction by a six-person jury did not violate the Constitution.

Social scientists interpreted the Court's empirical inquiry about what size jury best promotes group deliberation as a clarion call for information on the effects of size on jury functioning. After *Williams*, social scientists conducted substantial research comparing six-person juries with twelve-person juries. Justice Blackmun did not disappoint these researchers when, eight years later, the issue of jury size again came before the Court. In *Ballew*, the defendant was convicted by a five-person Florida jury of distributing obscene materials, a misdemeanor. In striking down the conviction, Blackmun surveyed the multitude of studies conducted since *Williams* in an opinion that has been likened to a social science article. Blackmun found that "these studies . . . lead us to conclude that the purpose and functioning of the jury in a criminal trial is seriously impaired, and to a constitutional degree, by a reduction in size to below six members."[61]

The fact of the matter, however, is that virtually all the research conducted between *Williams* and *Ballew* compared six-person panels to twelve-person panels. The research offered little or nothing to assist in making the choice between five and six. As Professor David Kaye, a prominent critic, suggested, Blackmun's "treatment of the statistical literature is, at best, careless."[62] Kaye suggests that *Ballew*'s reaffirmation of *Williams* represents "judicial intransigence—a willful disregard or cynical distortion of the writings of social scientists."[63]

Kaye's trenchant criticisms assume that *Williams* was or should have been on the table for reconsideration. But Blackmun operated from an altogether different premise. The constitutional question presented in *Ballew* concerned where to draw the line *given* the Court's earlier decision to uphold panels of fewer than twelve members. The empirical research available to the Court had no real relevance to this question—none of the studies compared five-person panels to six-person panels. Although Blackmun cast the social science research as responsive to the main issue, in reality, he used it to answer a different question than the one the scientists had researched. Blackmun "interpreted" the studies against the backdrop of the Court's decision in *Williams*. These new facts had to be integrated into the existing constitutional mosaic, a strategy that required reconciling the research with other principles of constitutional theory, most notably precedent. The "facts" became part of the interpretive reality of the Court, a process separated from any empirical reality.

Ballew reflects a fundamental conflict between science and law, a conflict that transcends the constitutional domain. Courts have relatively little latitude in choosing what questions to decide or in choosing the quality of supporting evidence available, but they must be expedient in providing answers. In the law, an answer based on incomplete information is often better than no answer at all. In contrast, scientists set their own agendas and have the luxury of seeking truth unconstrained by the demands of court dockets. Scientific explanations based on incomplete information are rarely better than withholding judgment pending more information. Still, many scientists set their agendas on the basis of questions the courts ask. This was typical of the research conducted after *Williams*. The Court, however, had already "answered" the question of the constitutionality of six-person panels and now addressed the constitutionality of five-person panels. The social scientists arrived too late.

The conflicts inherent in the methods of law versus the methods of science do not necessarily render the two disciplines incompatible. These conflicts do, however, exacerbate the Court's tendency to "interpret" facts in ways convenient to its general jurisprudence. Although requiring more courage, the Court should directly confront the challenges of integrating science into its decisions. In *Ballew* this might have meant Blackmun's frank statement that the costs associated with six-person panels were not of sufficient constitutional concern. While this might have led the APA to give their award to someone else, it would have been a more honest statement of the basis for the decision.

The greater concern, perhaps, with the basic conflict inherent between law and science is that the justices will give in to their ignorance of the scientific method. If Justice Blackmun did not entirely deserve the APA's award for distinction, Justice Powell, concurring in *Ballew*, deserves an APA award for disrepute. Powell agreed that the line should be drawn at six but criticized Blackmun's belief that social science might assist in the determination. Powell wrote:

> I have reservations as to the wisdom—as well as the necessity—of Mr. Justice Blackmun's heavy reliance on numerology derived from statistical studies.

Powell's likening of research based on statistics to numerology is like mistaking astronomy for astrology. It also might be exhibit one in making the case that lawyers and judges should receive training in the scientific method.

Very Interesting . . . , but Stupid

Exhibit two in a case for science education for judges and lawyers, and closely vying for the primary honors, is Justice White's opinion in a capital sentencing case, *Barefoot* v. *Estelle*.[64] *Barefoot* concerned the matter of predicting violence, an issue that, as we have seen, arises in a wide variety of legal contexts ranging from bail and probation decisions to civil commitment and capital sentencing hearings. Thomas Barefoot had been convicted of killing a police officer after being stopped for questioning in an arson investigation. Under Texas law, a murderer could receive the penalty of death only if the jury found that the murder had been committed deliberately and that "there is a probability that the defendant would commit criminal acts of violence that would constitute a continuing threat to society."[65] Based on the evidence of the crime and the expert testimony of two state psychiatrists, the jury found that the killing was deliberate and that, if allowed to live, Barefoot would be violent again.

Barefoot challenged his conviction on the basis that psychiatrists and psychologists cannot "predict with an acceptable degree of reliability that a particular criminal will commit other crimes in the future." At his sentencing hearing, the two state psychiatrists testified that Barefoot was a criminal sociopath. One of the experts, Dr. Grigson, popularly known around Texas as "Dr. Death," commented that there was a "one-hundred percent and absolute" chance that he would be violent in the future. Neither of the state's witnesses interviewed or even met Barefoot.

In its cross-examination of the state's experts, defense counsel asked whether they were aware of the research indicating that experts like themselves were not particularly able to predict violence. Although somewhat familiar with this work, both dismissed its relevance to them. Dr. Grigson argued that the belief that psychiatric predictions of violence are unreliable was accepted by only "a small minority group" of psychiatrists; "it's not the American Psychiatric Association that believes that," he pointed out.

The American Psychiatric Association, however, submitted a brief to the Supreme Court in *Barefoot* stating that they did believe it. The brief reported a multitude of studies that estimate that two out of three predictions of long-term dangerousness are wrong. Researchers have likened psychiatric predictions of violence to flipping coins in the courtroom. This metaphor understates their accuracy, though not by very much.

Justice White, writing for the Court, refused to be persuaded by the research indicating the inevitable fallibility of psychiatric predictions of violence. White's opinion shows a profound confusion regarding the scientific argument. This confusion, however, closely resembles Justice Blackmun's in *Ballew* and is attributable to White's need to contend with data that ran counter to the Court's precedent. As White remarked, "the suggestion that no psychiatric testimony may be presented with respect to a defendant's future dangerousness is somewhat like asking us to disinvent the wheel." Yet while the Court was disinclined to disinvent the wheel, it used a kind of formal logic that predated the wheel's invention.

White advanced the following argument to reject the data:

> If the likelihood of a defendant's committing further crimes is a constitutionally acceptable criterion for imposing the death penalty, which it is, *Jurek* v. *Texas*, and if it is not impossible for even a lay person sensibly to arrive at that conclusion, it makes little sense, if any, to submit that psychiatrists, out of the entire universe of persons who might have an opinion on the issue, would know so little about the subject that they should not be permitted to testify.[66]

The logic of White's syllogism is irreproachable:

1. *Jurek* holds that a dangerousness determination by a lay person is a "constitutionally acceptable criterion for imposing the death penalty."
2. All psychiatrists are also lay persons.
3. *Therefore*, a dangerous determination by a psychiatrist is a "constitutionally acceptable criterion for imposing the death penalty."

There are, however, profound problems with the soundness of White's syllogism, the least of which might be the continuing validity of *Jurek*. *Jurek* can easily be distinguished from *Barefoot* on the basis of the peculiar treatment accorded expert testimony in the trial process. Ordinarily, experts are not given carte blanche to testify to any fact that is within the competence of the lay public. Indeed, experts are generally defined as expert by their ability to assist jurors with facts not generally known. After all, experts are supposed to be . . . well, experts.

What is particularly instructive about the decision is White's struggle to integrate the data on psychiatric predictions of violence with the fabric of settled constitutional doctrine. White perceived the data in the same way he might contemplate other potentially inconsistent facets of constitutional interpretation, such as prior precedent or contemporary values. The data had to be "interpreted" in light of accepted doctrine. White's unwillingness to accept the radical alternative of overruling precedent and his inability to declare the fact in issue irrelevant led him to observe that despite psychiatrists' significant error rate, the adversarial process provides the means by which jurors can decide which psychiatrists are correct. Concerning this belief, White made the remarkable assertion, "Neither [Barefoot] nor the Association suggests that psychiatrists are always wrong with respect to future dangerousness, only most of the time."[67] As Justice Blackmun pointed out in dissent, this observation "misses the point completely," for "one can only wonder how juries are to separate valid from invalid expert opinions when the 'experts' themselves are so obviously unable to do so."[68]

As *Brown, Ballew,* and *Barefoot* illustrate, the modern Court does not differ from its predecessor Courts in *Lochner* and *Marbury* in using facts as an interpretive vehicle for constitutional adjudication. In this view, constitutional fact-finding, like the text, original intent (history), precedent, scholarship, and contemporary values, is something that must be integrated and shaped into the greater fabric of constitutional doctrine. Fact-finding has another thing in common with these other interpretive authorities: the clearer and more definite the authority, the more difficult it is for the Court to ignore it. For instance, no one seriously believes that the Court could freely interpret the constitutional mandate that the president be at least thirty-five years of age. Whereas equal protection, due process, and even free speech permit the Court much flexibility, the Court could only ignore a specific age requirement at the risk of losing its legitimacy.

The power of the Supreme Court depends on this legitimacy. As the only unelected branch of the federal government, the judiciary must constantly reflect on the basis for its power. Alexander Hamilton observed in *The Federalist No. 78* that the judiciary has "neither Force nor Will, but merely Judgment." The Court's efficacy depends on the power of its reasons for deciding a case. Still, although the Court is acutely aware of this lesson, it regularly exhibits much maneuverability when reading the Constitution. But discretion is a necessary part of every judge's job description, and any Constitution written to remove all discretion would certainly fail. The importance of restraining principles, such as text and precedent, lies in their cabining that discretion, not eliminating it. The Court retains legitimacy only as long as it remains within accepted bounds when exercising its discretion. Science offers much potential as an enforcer of those boundaries.

But You Can't Say *That*: Science as a Restraining Force

Sources of interpretation restrain the Court to the extent that they provide a degree of external control over the Court's proclamations. Hence, we can all read the Constitution as proscribing Congress from "abridging the freedom of speech." When Congress passed a law banning the burning of American flags, the Court considered itself bound to invalidate the law, since Congress meant to censor the message such protests made. To be sure, the Court could have upheld the law, and four justices argued for just this result. But to do so, the Court would have had to explain in some detail either why burning flags was not speech (and all of the justices found it to be speech) or why the Congress was nonetheless justified in banning speech, despite the First Amendment's clear statement otherwise. The text thus both guides and restrains the Court's decisions—but it almost never dictates the result. And the availability of the text for us—not to mention the president, Congress, state governors, state lawmakers and so on—to read restrains the Court substantially. Without the force of armies or the power over the purse, the Court relies entirely on its pronouncements being persuasive to the greater political community.

To demonstrate the restraining effect of empirical data, it need not be shown that the Court consistently heeds that data. Given the Court's history, such a showing would not be possible in any event. The value of science as a check on discretion is usefully assessed in comparison to the traditional sources of constitutional authority. Sources of authority restrain fairly modestly, typically by establishing the grounds for debate and the boundaries beyond which the Court may not venture. Much interpretive leeway remains.

The Court, of course, is often accused of mishandling the text, the framers' original intent, precedent, constitutional scholarship, and contemporary values. Constitutional scholars and pundits depend on such alleged excesses for their livelihoods. The critical response that routinely follows these alleged errors surely influences the Court. The justices read their reviews, both scholarly and in the general press. Recently, for instance, this phenomenon has been labeled the "Greenhouse effect," after the *New York Times*'s influential reporter Linda Greenhouse. In a very real sense, the public and critical scholars carry on a dialogue with the Court. Although in any one opinion the Court obviously cannot satisfy all its critics, critical commentary undoubtedly influences subsequent cases. The Court's ignorance of or disdain for science similarly leads to scholarly attempts to educate the justices, which in some measure influence their later decisions. In this way, science functions like the other sources of interpretation in shaping the contours of the Constitution.

Whenever the justices misuse empirical research, they become the subjects of significant criticism. In fact, such critical review on a subject the justices find difficult to master probably has made them reluctant to delve into the niceties of research methodology at any time. Justice Brennan demonstrated this reluctance in *Craig v. Boren*.[69] *Craig* involved a challenge to a state law that prohibited men under twenty-one years of age from purchasing "nonintoxicating" 3.2 percent beer while permitting women over eighteen years of age to buy it. The law was challenged under the Equal Protection Clause of the Fourteenth Amendment for discriminating on the basis of gender. Under the heightened scrutiny of the law the Court employed, Oklahoma was obligated to justify the discrimination by showing that the law "serve[s] important governmental objectives and [is] substantially related to achievement of those objectives." Oklahoma justified the discrimination on the basis of statistical studies indicating that young men account for a disproportionate share of drivers arrested for driving while intoxicated. Justice Brennan began his analysis by criticizing the research methodology of the studies and doubting their value. And, in fact, the studies produced by Oklahoma were methodologically weak and proved little. Justice Brennan was correct to question their empirical significance. He went on, however, to cover his empirical flanks by trying to distance his constitutional analysis from a methodological critique about which he was not confident:

> There is no reason to belabor this line of analysis. It is unrealistic to expect either members of the judiciary or state officials to be well versed in the rigors of experimental or statistical technique. But this merely illustrates that proving broad sociological propositions is a dubious business, and one that inevitably is in tension with the normative philosophy that underlies the Equal Protection Clause.[70]

This statement displays profound ignorance about the connection between science and constitutional law. How can the Supreme Court possibly scrutinize whether a state law "is substantially related to" an "important governmental objective" unless it understands the empirical argument forwarded by the state? The *Craig* Court invalidated a law passed by the democratically elected representatives of Oklahoma by rejecting a justification that, by Brennan's own admission, he did not understand. This irresponsible statement, however, comes from Brennan's reluctance to make definitive assertions on a subject that he is not "well versed in."

Escaping responsibility for knowledge of science is not so simple for the justices. The Court's own jurisprudence makes plain that they must become "well versed in the rigors of experimental or statistical technique." As Oliver Wendell Holmes understood more than one hundred years ago, "For the rational study of the law the black letter man may be the man of the present, but the man of the future is the man of statistics and the master of economics."[71] Holmes has been proved correct. Too much of the Court's docket requires a knowledge of empirical techniques for the Court to continue pleading *non sum informatus*.

Ultimately, persistent misapplication of empirical data undermines the Court's legitimacy. Rulings that rest on faulty premises have little or no persuasiveness, for they lack rationality and judgment—the source of judicial power. Although perhaps only implicitly, the Court seems to fully appreciate this lesson in the context of scientific premises. Hence, when the Court is confronted by a strong empirical argument that is contrary to the conclusion they seek to reach, they go to great lengths to avoid the science. Just as in the creation science case cited in Chapter I, *Edwards* v. *Aguillard*, the Court would prefer to stay on the ground of abstract legal principles it knows so well. But even in these cases when the Court does back flips to avoid the implications of the factual context, scientific research can have a profound impact on the constitutional dialogue.

Science Will Not Be Ignored

McCleskey v. *Kemp*[72] is a case that is both troubling and promising in the lessons it offers for the law and science connection. Warren McCleskey, a black man, was sentenced to death for the killing of a white police officer during the course of a robbery in Fulton County, Georgia. McCleskey introduced at trial an extensive and sophisticated study conducted by Professor David Baldus and others (the "Baldus study") indicating, among other things, that "defendants charged with killing white victims were 4.3 times as likely to receive a death sentence as defendants charged with killing blacks."[73] The study indicated, therefore, that juries considered crimes against whites to be more egregious than crimes against blacks and imposed punishments accordingly. McCleskey argued that the Baldus study demonstrated that his death sentence violated the Equal Protection Clause of the Fourteenth Amendment and the Eighth Amendment's ban on cruel and unusual punishment. I will concentrate on the Court's response to McCleskey's Eighth Amendment claim.

Perhaps surprisingly, the Supreme Court *assumed*, for the purpose of its opinion, that the Baldus study was valid. Nonetheless, and less surprising, it concluded that McCleskey's claim still failed. Justice Powell, whose view of statistical studies as modern numerology we met above, wrote the opinion for the Court.

Prior to *McCleskey*, the Court had ruled that the death penalty could "not be imposed under sentencing procedures that create a substantial *risk* that the punishment will be inflicted in an arbitrary and capricious manner."[74] The condemned does not have to prove that race affected his sentencing decision; the Eighth Amendment's concern is the "sentencing system as a whole."[75] Hence, the condemned establishes a constitutional violation by demonstrating a "pattern of arbitrary and capricious sentencing." This systemwide standard is especially well suited to statistical proof.

Although not free of methodological error (as is true of all research), the Baldus study appeared to document just such a "pattern of arbitrary and capricious sentencing." Justice Brennan, apparently a born-again convert to the value of statistical studies, demonstrated a newfound ability to understand them. He summarized some of the inferences to be drawn from the Baldus study in a dissenting opinion:

> For the Georgia system as a whole, race accounts for a six percentage point difference in the rate at which capital punishment is imposed. Since death is imposed in 11% of all white-victim cases, the rate in comparably aggravated black-victim cases is 5%. The rate of capital sentencing in a white-victim case is thus 120% greater than the rate in a black-victim case. Put another way, over half—55%—of defendants in white-victim crimes in Georgia would not have been sentenced to die if their victims had been black.

Under the existing "arbitrary and capricious manner" test, Brennan's analysis of the Baldus study should have justified a ruling in favor of McCleskey. The *McCleskey* Court, however, changed the legal standard in order to avoid the conclusion compelled by the Baldus study. Whereas prior to *McCleskey* a petitioner needed only to show a risk of discrimination in the system as a whole, the Court now held that the Eighth Amendment required a particularized showing of discrimination in the petitioners' own case. This new legal standard rendered the statistical proof irrelevant: "Even Professor Baldus does not contend that his statistics *prove* . . . that race was a factor in McCleskey's particular case." This view led Powell to comment, "At most, the Baldus study indicates a discrepancy that appears to correlate with race."

Ironically, Powell's statement that despite being infected with racial bias the Georgia capital sentencing system is constitutional illustrates the power of science as a "restraining principle" of constitutional interpretation. The available data forced the Court to confront the value choices it was making, thus rendering the decision more difficult and more controversial than it would have been if the Baldus study never existed. The study forced the Court to admit that it was upholding McCleskey's sentence despite the fact that systemic prejudice was present in the Georgia capital sentencing system. To be sure, with or without the Baldus data, the Court intended to uphold McCleskey's sentence. But the public would never have known that the Court was willing to uphold a defendant's sentence in the face of systemic prejudice without the Baldus study. In this way the Court was held accountable for its decision. Moreover, it alerted the political community to the true basis of the Court's decision, thus permitting groups to mobilize and respond to the empirical realities in Georgia. *McCleskey* was met by a barrage of criticism from media critics and academics, as well as a flurry of proposals to reverse its effects through legislation introduced in chambers from Athens to Washington, D.C.

This is the lesson of *McCleskey*. Without a sound appreciation of science, the Court could do little but accept the conclusions of the Baldus study. And there is little question that Powell did not understand the science. As he himself admitted after the decision, "My understanding of statistical analysis . . . ranges from limited to zero."[76] If they had understood the science, the justices might have paused and reflected on the racial realities of the system Georgia used to execute people. Instead, in order to avoid the legal effect of these realities, the Court was forced to reconstruct the law itself, thus rendering the statistical proof irrelevant. This reconstruction, however, was poorly crafted and feeble. It was immediately subjected to the strong winds of public condemnation. It is a construction unlikely to stand the test of time.

A Scientifically Sophisticated Constitutional Jurisprudence

Throughout its history, the Supreme Court has advanced certain facts that, in actuality, substituted for value judgments or were convenient claims about the empirical world asserted to support constitutional conclusions reached on other grounds. In *Marbury* v. *Madison*, the Court relied on the "fact" that legislators are more likely than judges to deviate from constitutional confines to justify Marshall's judgment that the judiciary should have the power to review the acts of the legislative and executive branches of government. In *Brown* v. *Board of Education*, the Court relied on the fact that segregation promoted feelings of inferiority in school children to support its determination that separate but equal facilities could no longer be constitutionally tolerated. In *Roe* v. *Wade*, the Court relied on medical science as a convenient proxy by which to strike a balance between the conflicting absolutes of a woman's right of choice in matters affecting her body and the state's legitimate interest in protecting the potential life of the fetus. As these cases demonstrate, the Court sometimes finds it expedient to invoke factual suppositions in lieu of declaring its underlying value judgments. Also, as these cases and others illustrate, the Court has not hesitated to jettison inconvenient facts as they became known through scientific research, either by confessing ignorance of the scientific method, ignoring the true import of the research, or changing the legal rule to render the research irrelevant.

My recommendation for improving the Court's use of facts is elementary in theory, however ambitious it might be in application. Simply put, the Court must begin to take factual statements more seriously. This deceptively simple proposal contemplates a Court with a deep appreciation for the essential core and subtle margins of the scientific method. It does not mean that research results will dictate constitutional meaning or that the Court will not sometimes reconsider its jurisprudence in light of subsequent developments. It does mean that the Court will integrate knowledge of constitutional facts into its jurisprudence fully in light of the empirical worth of the research rather than the rhetorical benefit it lends to a conclusion reached on other grounds. Under my proposal, the necessity of integrating constitutional facts into the complex mosaic of constitutional law would not change. Indeed, quite the contrary is true. What I suggest is that constitutional facts be given full membership in the family of authorities the Court ordinarily relies upon to give the Constitution meaning.

When the Supreme Court examines prior case law and identifies the holdings of precedent, the public legitimately expects that it matters to the resolution of the case. For example, in *Roe v. Wade*, the Court considered whether prior cases had defined a fundamental right of privacy in the Constitution. The Court had identified such a right in several cases, including *Griswold v. Connecticut* and *Eisenstadt v. Baird*. In *Griswold*, the Court found that the right of privacy necessitated invalidating a Connecticut law that prohibited the use of contraceptives by married couples. In *Eisenstadt*, the Court extended *Griswold* and invalidated a Massachusetts law prohibiting the distribution of contraceptives to unmarried individuals. In 1973, when *Roe* came before the Court, a central issue concerned the definition of the parameters of the right of privacy articulated in these earlier cases. How extensive a right is the right of privacy? And although it clearly extends to an individual's decision to use contraceptives, can it be understood to extend to the sort of reproductive choice represented by an abortion decision?

Most constitutional commentators agree that the Court in *Roe* substantially expanded the previous scope of the right of privacy. This is no criticism, since the Court regularly adjusts its constitutional jurisprudence over time. However, the *Roe* Court was obligated to fit its 1973 understanding of privacy into the existing precedential fabric of constitutional law or explain its failure to do so. Although the Court frequently fails to follow precedent, it does so with some trepidation and usually with awareness of the costs associated with any departure from settled law or settled expectations.

Fact-finding should occupy a similar position in the processes of constitutional interpretation and adjudication. In saying what the Constitution means or how it applies, the Court cannot be permitted the latitude of ignoring facts if or when they become inconvenient. For example, if segregated education turns out to lead to better outcomes, however this notion is defined, then the Court should be expected to respond to this result. The Court need not overrule its holding in *Brown*, since this fact might still not control. It might be, for instance, that the Court reads the Equal Protection Clause to mandate an integrated society (at least as far as state action is concerned) and that today the factual question of 1954 is no longer relevant. But the Court cannot simply assert that segregated education is "inherently unequal" if research indicates that the contrary is true. The Supreme Court is indeed powerful, but it is not powerful enough to put the earth at the center of the universe for the sake of constitutional doctrine.

The simplest way to determine whether constitutional facts are being taken seriously by the court is to ask, "What happens if the facts change?" We know what occurred after *Brown* when the argument was made that the facts had changed: nothing. The *Brown* court did not take the social science research seriously. *Roe* v. *Wade* is another example in which the court relied on facts that were not taken very seriously. Viability, as we saw above, was a political compromise between the fundamental right of women to control their bodies and the compelling interest of governments in protecting "potential life." The medical fact of "viability" conveniently occurs at around twenty-four weeks, a perfectly understandable point at which to strike a compromise. But no one could seriously believe that if medical technology moved viability to the sixth week of pregnancy that the Court would follow suit. The Court simply never took the medical science as a determinant of a constitutional right that seriously.

The advantage—for both the Court and Court watchers—of asking "What happens if the facts change?" is that the true bases of the decision become apparent. The true basis for *Brown* was that the Court had reached a moral judgment about segregation before much of the nation had. The asserted deleterious effects of segregation were convenient fodder to support that judgment. The true basis for *Roe* was a political compromise between two absolute and irreconcilable principles, and the medical fact of viability was a convenient tool to mask the apparently arbitrary point selected to reconcile this conflict. In both *Brown* and *Roe*, science offered a seemingly objective source of authority to lend legitimacy to decisions reached on other grounds. But ultimately, it was an investment in short-term legitimacy purchased at the expense of the Court's long-term legitimacy.

Foundations of Legitimacy

The Supreme Court's role in American constitutional democracy is in many ways a precarious one. The Court has no power over the purse or the sword; its power instead resides merely in the persuasive force of its judgment. The Court has always guarded this power base with jealousy and with special solicitude for how its pronouncements would be received by the greater community. This care has resulted in an institution that is deeply respected by the larger society. Yet it is not a blind reverence, for "we the people" still retain ultimate authority over all matters that come before the Court. In evaluating the Court's performance, the larger community considers not just the outcomes of particular cases but also the manner in which the Court reaches these results. We expect judges to be guided by the past but not to be prisoners of it; and we expect them to be wise in anticipating the future but not to lead social revolutions to produce it.

In the day-to-day task of saying "what the law is," the Court regularly consults authorities beyond the text in order to effectuate its commitment to the past and its obligations to the future. The Court traditionally considers original intent, precedent, scholarship, and contemporary values for assistance in determining constitutional meaning. The Court also routinely relies on facts susceptible to scientific methods of testing. These facts, like other constitutional authorities, do not dictate particular outcomes. They merely guide and restrain the Court. In this way, they contribute to the dialogue called constitutional law.

The quality of this dialogue, of course, depends on all the participants fully understanding the language and terms used in the discussion. The Court continues its adherence to ignorance at the peril of substantial losses in its legitimacy. Imagine public reaction to the Court's confession of ignorance of history and historical methods when interpreting the Constitution. Yet the Court seems to hardly hesitate in professing its scientific and statistical illiteracy. But science is no less necessary to constitutional interpretation than history. The Court's judgment, and hence its power, depends on its effective employment of scientific research.

IT'S NOT JUST A BAD IDEA, IT'S THE LAW

Science in the Legislative Process

No man should see how laws or sausages are made.
— OTTO VON BISMARCK

It could probably be shown by facts and figures that there is no distinctly native American criminal class except Congress.

— MARK TWAIN

The public's perception of politicians is decidedly negative. It is the nature of democracy, of course, that legislators are invariably criticized, for all political choices made, or avoided, because some group inevitably is on the losing end of the decision. As Harry Truman observed, presciently in his own case, "A statesman is a politician who's been dead ten or fifteen years."[1]

It is instructive to look back at the founding generation, a generation that seemed to produce so many great statesmen, to obtain some perspective on our current group of legislators. One fact that appears consistent between the two groups is that the vast majority of the leaders of 1776 and those of today were trained in the law. This should be neither surprising nor necessarily disturbing, since legislators, by definition, are "lawmakers." A solid grounding in the law would appear essential, at least for some percentage of legislators. A major difference (no doubt among many differences), however, between the leaders of the Revolution and those of our own time is their knowledge also of the science of their day. Many of the major political figures of the late eighteenth century had either a solid grounding in science or were themselves scientists. Benjamin Franklin, for example, was admitted into London's prestigious Royal Society of Scientists, and Thomas Jefferson was, well, Thomas Jefferson. James Madison, James Monroe, John Adams, and even George Washington were extensively trained in scientific or engineering principles.[2] In comparison, our present group of legislators is notable for their lack of science training. Of the 535 members of the United States Congress, fewer than 1 percent have any significant training in science. Today, the route to being a "statesman" is through law or business. Although science and technology have become immeasurably more important to modern affairs of state, our leaders' knowledge of these subjects has dwindled to near zero. Our founding generation built a political model based partly on an analogy to Newtonian mechanics. Our current generation presides over this political system with little knowledge of Newton, much less of Einstein, Bohr, or Feynman.

Any overview of how legislators use science must begin with some sense of how they make decisions generally. In general, the motives driving legislative politics can be described along three basic dimensions: for hire, poll following, and enlightened republicanism. Politicians embody these dimensions to varying degrees at various times.

The cynical, for-hire view of politics holds that legislators are bought and paid for by special interests. The main preoccupation of most legislators, according to this view, is reelection. The political cynic would see science as something that is manipulated or shaped for rhetorical advantage by legislators desirous of using science for selfish ends. An alternate cynical description of politics contemplates legislators who are mere proxies for majority sentiment. But even this more optimistic view of politicians' motivations seems to expect little scientific sophistication from legislators. Since the voters crunch the data and make the important choices, the representatives need only follow directions. The third view of politics, the enlightened republican view, sees legislators as representatives who combine their good judgment with the voters' expressed wishes and, through reasoned deliberation, choose a course to follow. This is Jimmy Stewart in *Mr. Smith Goes to Washington*, both a man of the people and a man who leads the people. Science, in this hopelessly hopeful vision, informs public debate but does not drive it. Science would thus be fully integrated into the great question debates that preoccupy our public representatives.

None of these three caricatures of the legislative process is entirely accurate. Legislators are often beholden to special interests, but only rarely is a special interest group powerful enough to materially control or completely condemn legislation. Historically, the NRA and the tobacco lobby might have had this kind of power. Even these powerful forces, however, have lost much—though far from all—of their influence. The realities of the legislative process have in fact changed little since the nation was founded. Much of the Madisonian genius embedded in the American political system lies in the institutional checks on special interests that Madison called factions. These checks result in an inherently conservative process in which it is relatively easy to block legislation but enormously difficult to enact it. Although special interest money or political support might influence many political decisions, it rarely, if ever, drives them entirely.

Similarly, legislators who wish to be reelected surely remain aware of majority sentiment, as the vast increase in reliance on polling data by incumbents demonstrates. But most issues before Congress, especially ones involving complex science, do not attract straightforward majority support or condemnation. Instead, depending on how the question is phrased, the public tends to be divided, if not ambivalent, on many of the issues, even many of the great issues, before Congress at any given time. If any one generalization might be made about the American public today, it is that they want their cake and they want to eat it too. Thus, Americans want to explore Mars but not spend billions of dollars building the spaceships needed to get there; they want to save spotted owls but do not want to lose the logging jobs that would be eliminated if the owls' habitat is protected. Congress, therefore, confronts a divided and fickle public when it weighs the costs and anticipates the opportunities science presents.

Finally, few lawmakers today bear even a passing resemblance to Jimmy Stewart. Politics is too often a selfish business, and there is little hope that detached and intelligent deliberation will soon become the leitmotif of congressional debate. As Mark Twain summed up the popular sentiment, "Suppose you were an idiot. And suppose you were a member of Congress. But I repeat myself." Still, many and perhaps most legislators are committed and principled people who desire to do good for America, at least as they define good. Any study of the legislative process that ignores the high ideals of lawmakers would be incomplete. This lesson is especially so for a study on legislators' use of science, since from abortion regulation to zoo funding, science policy so often implicates the profound issues of our time. In the final analysis, the art of legislating involves and requires an understanding of what is good for scientists, what is good for citizens, and what is good for legislators.

Any viable theory of legislation, then, must take into account the many contradictions and conflicting forces impinging on the democratic process. Empirical truth is just one of these forces. Nonetheless, legislators speak as though empirical truth is an essential object of their attention. Whether it is the health effects of air pollution, how employment figures affect the inflation rate, or the factors that cause crime rate fluctuations, legislators constantly speak in scientific terms. We might be tempted to conclude that representatives know something about science or would care to learn. We should avoid the temptation.

As in other areas of the law, legislators often use a facade of science to legitimate decisions—decisions not always made in reliance on or even consistent with the science being cited. Legislators, however, get away with scientific legerdemain far easier than judges. There is a simple reason for this. Judges typically write opinions explaining their decisions, which are then published and pored over by legions of lawyers, scholars, students, and other judges. Legislators have no similar tradition of explaining the reasons for their decisions. To be sure, laws are passed accompanied by forests of committee reports and transcripts of legislative hearings. But these reports neither specifically set forth arguments for the decision nor garner the kind of attention trained on judicial decisions. Because they are not held accountable for their knowledge, legislators feel little pressure to truly deal with the complexities of science. Only the conclusion counts or, perhaps, the sixty-second sound bite the conclusion generates. The reasoning or principles that underlie the conclusion are of minor legislative concern.

When it comes to learning about science or any technical subject, legislators are not so much interested in wrestling with the details as in giving various constituencies an opportunity to be heard. Legislators love to hold hearings that seemingly promise much educative value. Experts regularly testify before legislators in much the same way they do before judges making admissibility decisions. In fact, legislative hearings are more luxurious than this, sometimes resembling a classroom more than a hearing room. But any expert who has appeared before both legislators and judges can testify to the differences. Legislators want to put on a show, while judges want to be educated. For legislators, the important thing is that the issues be aired publicly; for judges the important thing is that the issues be understood.

Legislators interested in being noticed or receiving a television moment also have a built-in disincentive to hearing or speaking too much about science. Science is complex, general, and diffuse and involves too many cold numbers. Although the adventure of science surely titillates layman and legislator alike, most Americans find the details of science to be extremely difficult, excessively tedious, and assiduously soporific. In this age of Oprah and Geraldo, personal anecdote is the way to be noticed.

While Congress has an overwhelming influence on the practice and pursuit of science in the United States and constantly relies on it, as an institution it exhibits a shocking lack of curiosity about the subject. Probably the best illustration of legislative science deficiency is Congress's decision on September 30, 1995, to abolish the Office of Technology Assessment (OTA).

Goodbye OTA, Hello Ignorance

Congress's decision to close the doors of the Office of Technology Assessment is comparable to a deaf person blinding himself in order to save money on eyeglasses. Established in 1972, the OTA provided comprehensive, in-depth reports to Congress on a wide range of scientific issues, including medical research, climate change, the space program, telecommunications policy, and nuclear weapons management. In addition, the OTA regularly evaluated and reported on alternative policy choices. It operated essentially free of political influence and was controlled by a bipartisan board of directors. The OTA provided Congress with straightforward and detailed research on matters of great scientific and technological import for policy making. Congress killed it to save money. Fittingly, Congress did not ask for or provide an empirical assessment of the cost-effectiveness of terminating the OTA.

The OTA's $22-million budget was less than 1 percent of a total annual congressional budget that exceeds $1.5 trillion a year. Ironically, its small size and political neutrality made it particularly vulnerable to budget cuts. Representative Amory Houghton offered a playground analogy: "You don't cut the big bully down to size because he's too big to handle. But the little guy, who may even be the next genius, you can pummel the dickens out of him. And that's what happened to OTA."[3] Moreover, the OTA's political neutrality made it a "little guy" with no big friends to defend it. Dr. John H. Gibbons, then head of the White House Office of Science and Technology, stated the point succinctly: "If you belong to everyone, you belong to no one."[4]

According to some congressional opponents of the OTA, scientific and technological information will now be obtained from other sources, both public and private. Senator Connie Mack argued, for instance, that "the explosion in technology in America has been accompanied by an explosion in information on technology."[5] Where this information will come from and what form it will take remain unanswered questions. In the public sector, many opponents cited the Congressional Research Service (CRS) as a substitute resource. But Jane Griffin, director of the CRS's science division, demurred: "The OTA did major long-term studies, which involved a large number of outside groups. It's not something we have the resources to do."[6] In the private sector, presumably corporations and their lobbyists will be relied on for the kinds of assessments once done by the OTA. This is the sort of "fox guarding the chicken house" legislative policy making that could have come directly from Philip Morris or R. J. Reynolds.

Legislators' inability to understand what the OTA did might have contributed to its demise. As we know, most legislators are not distinguished by their great knowledge of math and science. Additionally, legislators are inclined to look for the bottom line or conclusion, thus rendering the premises somewhat irrelevant. Richard Nicholson, executive director of the American Association for the Advancement of Science, noted: "There used to be a time when knowledge was power. Now it seems Congress has decided it's a nuisance."[7] Alan Crane, a nuclear proliferation analyst, echoed this view: "If you have all the answers you want, then you don't need analysis."[8] The OTA provided Congress with more detail and complexity then the legislators needed—or could handle—to make decisions. Science sometimes gave answers that were inconsistent with legislators' policy predilections. Better to receive the bottom line from your policy supporters or to remain ignorant of the science altogether. Ignorance might not be "bliss," but it is politically expedient.

When Science Goes to Congress

Practical politics consists in ignoring facts.
— HENRY ADAMS

Congress, of course, cannot entirely avoid confronting the details of science, and individual legislators are far from reticent in speaking on a myriad of scientific issues. As the principal lawmaking body in the United States, Congress does substantial business with scientists, makes decisions that profoundly affect science, and legislates on a myriad of matters to which scientific research is relevant. In fact, Congress's scientific dealings can be divided into three basic types.

The first way in which Congress meets science is in the area of big-ticket purchases. Congress often funds major scientific projects, with some involving basic science while others have a more applied focus. I consider two salutary examples of big-ticket purchases from recent times, the superconducting supercollider (SSC) and the space station.

A second primary role for science in Congress is legislators' inclination to sometimes contribute to the debates—and occasionally end them—concerning the great moral or ethical issues that whirl around science itself.

One of the more curious examples of this moralizing, considered here, was the legislative debate that followed the news that Scottish scientists had successfully cloned an adult sheep.

Finally, Congress exercises its regulatory role in a wide assortment of contexts in which science-based arguments are central to the policy choices presented for decision. Most environmental questions, for example, are overtly of this type. In reality, virtually all legislation before Congress has some science or math underlying it, if only economic theory or simple cost projections. I use the story of the great saccharin scare of the 1970s to illustrate this regulatory aspect of science.

Big-Ticket Science

Congress spends colossal sums of money on both basic and applied science. Most of these dollars are funneled through administrative agencies directly involved in overseeing scientific research. Examples with which we are most familiar are the National Aeronautics and Space Administration (NASA), the Department of Energy (DOE), the National Institutes of Health (NIH), and the National Science Foundation (NSF). There are also sundry other agencies that fund or rely on scientific research, including the Nuclear Regulatory Agency (NRA), the Federal Aviation Administration (FAA), and the Environmental Protection Agency (EPA). These administrative agencies, and thus the executive branch of government, ultimately are responsible for implementing most scientific policy in the United States. Congress is generally very attentive, too, since it controls the purse strings. Nothing grabs congressional attention quite as effectively as the big-ticket items on scientists' shopping lists. Here I consider two recent congressional purchases, the superconducting supercollider and the space station. The first Congress decided to return to the store, after expenditures of over $2 billion; on the second, Congress still owes staggering payments in amounts ranging, depending on whom you listen to, from $17.4 billion to $94 billion.

A Supercolossal Mistake

By any measure, Congress' handling of the superconducting supercollider (SSC) was a debacle. For supporters of the supercollider, Congress's decision in 1993 to terminate the project signaled a retreat from America's preeminence in high-energy physics and illustrated a profound lack of commitment among lawmakers to basic science. For opponents of the SSC, the termination order came about three years and $2 billion too late. In the end, everyone had reason to complain. American taxpayers spent billions for a fourteen-mile hole in the ground outside Waxahachie, Texas, and an expensive lesson about what occurs when science and politics meet. It is an important lesson, though probably not worth the cost. This is especially so since it is a mistake Congress almost certainly is doomed to repeat.

In the early 1980s, a group of theoretical physicists proposed the construction of a supercollider to study matter at energy levels beyond what was available at existing collider facilities. The supercollider's principal purpose was to permit physicists to further explore and experimentally test theories of the fundamental forces of the universe. The goal, quite simply and profoundly, was to determine the fundamental nature of matter. Nobel laureate Richard Feynman characterized this goal as "the greatest adventure the human mind has ever begun." Senator Bennett Johnston, a strong supporter of the supercollider, explained, "Accelerators are to particle physics what telescopes are to astronomy, or microscopes to biology."[9] To both peer into the smallest elements of nature and to "see" the first few moments after the "big bang" that created our universe, physicists needed a bigger instrument. In terms of a microscope, the fifty-four-mile accelerator would resolve distances of less than a billionth of a billionth of a meter;[10] in terms of a telescope, it would observe the very farthest reaches of the universe or, stated more accurately, the very first moments after the universe came into being.

The proposed collider would have had an energy level of 20 trillion electron volts (TeV), twenty times greater than the most powerful collider then in existence. It would operate by accelerating and guiding two beams of protons in opposite directions around a ten-foot-wide track that was fifty-four miles in circumference. The proton beams would be guided by approximately 10,000 large superconducting magnets located around the track. The protons would collide at speeds close to the speed of light, and detectors would measure the resulting carnage for evidence and insights about the fundamental materials and laws of the universe.

There is an old *Peanuts* cartoon in which Linus and Charlie Brown are talking as they peer over a wall and stare into space. Linus says to Charlie Brown, "Can you believe that people used to think the earth was flat?" He and Charlie Brown then laugh uproariously for several panels. In the last panel, Charlie Brown, now looking perplexed, asks curiously, "What do we believe now?" The supercollider's principle purpose was to permit physicists to advance beyond the limitations inherent in the current framework of physics for understanding the basic constituents and forces of the universe, what is called the "standard model." To appreciate the value of the SSC, it is helpful to briefly summarize what we believe now about the fundamental forces and matter that make up our universe.

The standard model predicts that all matter is composed of four particles. The first two, protons and neutrons, are made up of two quarks and reside in the atomic nuclei. The third is electrons, which surround the nuclei. Finally, neutrinos, fast, virtually massless objects, are produced by nuclear reactions inside stars.

The standard model also posits the existence of four fundamental forces that affect matter: the weak nuclear force, the strong nuclear force, the electromagnetic force, and gravity. The weak nuclear force describes the force that holds electrons in their orbits around the nucleus and triggers radioactive decay; it also describes the transformation of electrons and protons into neutrons and neutrinos during nuclear reactions. The strong nuclear force is the glue that holds the nucleus together. The electromagnetic force builds atoms into molecules and binds molecules together to form the macroscopic objects, such as ourselves. Gravity operates at every scale, but because it is so weak, it is usually associated with large stellar bodies. Gravity's strength is proportional to the mass of the bodies involved. A driving motivation of contemporary physics has been to identify underlying simplicities that would permit unification of two or more of these fundamental forces. Theorists Steven Weinberg and Abdus Salam made major progress in unifying the weak nuclear and electromagnetic forces into a single "electroweak" force. Recent advances suggest that the strong nuclear force and the electroweak forces might soon be brought together into a "grand unified" theory, thus reducing the number of apparent forces to two. The ultimate holy grail of theoretical physics, however, is the "supergrand" unified theory, which would account for gravity as well and thereby bring all the forces of nature into a single unified framework.

In addition, a major preoccupation of physicists endeavoring to unify the different components of the standard model is the question "Where does mass come from?" The leading candidates for this distinction are Higgs particles, named after Peter Higgs, who first posited their existence. The basic theory is that space is awash in Higgs particles and a particle's interaction with this "Higgs" soup gives it mass. Hence, according to the Weinberg-Salam model, electrons and neutrinos are the same when no Higgs particles are around. When Higgs particles are present, either electrons or neutrinos become massive, depending on the type of Higgs particle in the soup du jour. An electron's mass thus comes from the chowder in which it swims, while a neutrino has little or no mass as it moves effortlessly through a light chicken broth.

Understandably, much of the congressional debate throughout the years focused on what the supercollider would produce or achieve. Legislators did not demand a guarantee that the SSC would provide a final theory of everything, but they sought some comfort that the project would not be a total bust. In large part, supporters and critics of the project generally concurred about the value of SSC for particle physics. Beyond providing insights that might assist in solving the limitations of the standard model, however, few seemed to agree about anything. Over time, many bold claims were made in the name of the SSC.

Virtually all scientists agreed that, though there was some risk that the supercollider would make no major discoveries, in all likelihood it would offer experimental tests and produce results that would substantially advance particle physics. In particular, the project would begin to unravel certain basic contradictions in the standard model and thus help move theorists toward the supergrand unified theory. It would help resolve the question, for instance, why mass exists and should provide experimental evidence (or refutation) for Higgs particles. Finally, and perhaps most important, it would extend the frontiers of physics beyond anything that could be imagined today. The supercollider thus was viewed as the stepping-stone into the next century when physical and cosmological imaginings such as multiuniverses and superstring theory might all yield to experimental test.

Over the years, however, as proponents became increasingly alarmed that the supercollider was endangered, the claims for the benefits likely to be derived from this multibillion dollar investment became increasingly heroic. Some supporters went so far as to claim that the collider would contribute to a cancer cure.[11] The reality, however, was that the spin-offs of the SSC were not only unknown but unknowable. It is true, as proponents argued, that all of the major breakthroughs in physics in the twentieth century have contributed to technical applications now integral to modern society. A few general examples regularly cited before congressional committees were semiconductors, dry photocopiers, and digital computers. More specifically, previous accelerators had contributed the electron synchrotron, which permits the construction of integrated circuits with very high circuit element densities and which have enormously important applications for medicine, chemistry, and material science. The supercollider almost certainly would have contributed many technical advances and spin-off products as well. But the issue was whether these would have been worth the price tag. And, if not, the question remained, is discovering the basic laws of the universe, and perhaps gaining additional insights about the origins of the universe, worth the cost? This, inescapably, was a political question.

Despite the purity of the scientific aspirations accompanying the supercollider, the politics of the project were as simple and base as politics everywhere. Two closely related political questions were posed by the super collider project. The first concerned the pork barrel politics of congressional spending and the perennial political query "What's in it for me?" The second question can be described as the pork barrel politics of science, and it occupied politicians and scientists alike: the choice between "big science" and "small science" and which scientific communities would benefit from the government's largesse. Both questions fall under the umbrella topic of special interests, though only sometimes did the participants in the debate admit their selfish motives.

Pork of the Political and Scientific Varieties

Political Pork

Funding for science is no different than funding for other major federal purchases. Legislators regularly gauge the value of a project by the number of jobs it creates in their state or district. Just as contracts for the B-1 bomber were spread around all fifty states, congressional expenditures on science are expected to reap similar political rewards. The supercollider largely ran afoul of this central premise of politics. Indeed, nearly without fail, whenever an opponent of the SSC complained about the costs, proponents would respond that the benefits were spread broadly throughout the nation. In the end, however, the benefits were not spread around evenly enough.

The big winner of the supercollider sweepstakes was Texas, which, though it had to pony-up a billion dollars of its own, would have received thousands of new jobs and a prodigious influx of capital, as well as the prestige associated with being the home to this project. But other states, too, would have been winners. Louisiana, for example, expected to build the 10,000 magnets needed to guide the proton beams. Not surprisingly, legislators from Texas and Louisiana were strong supporters of the SSC. One of the strongest and most powerful supporters was Louisiana's Senator Bennett Johnston, who, as chairman of the Senate Energy Committee, oversaw many of the hearings on the supercollider. When the possibility of bringing Japan into the project by having it supply some of the magnets was proposed, Johnston bridled at the thought. "Let Japan dig the ditch or string the wires or something," he suggested.[12]

A problem from the start was describing the supercollider in a way that would appeal to the purchasers of the behemoth. Senator Phil Gramm from Texas, a strong supporter of SSC, knew what the project was "really" about. He lamented, "It is too bad . . . that when people write about high-energy physics in the newspaper, they talk about the birth of the universe and the cosmos, things that people basically may not know much about and, quite frankly, may not care anything about."[13] For

Gramm, the value of high-energy physics came from jobs, national security, and American preeminence in the world. Because of science, he explained, "we are richer, freer, happier, more powerful today, and we exert a much greater control over our lives and future and the destiny of the world than we would exert had that science and technology not existed."[14]

Another special interest complaint of many legislators was that the university winners of the SSC sweepstakes were the usual dozen or so elite institutions that always came out on top. In the hearings, legislators repeatedly voiced their concern that the smaller institutions and state schools, which teach most of the physics students in the nation, were not benefiting from federal spending. The question of academic elitism is closely related to the issue of big science versus small science, which most legislators ultimately identified as *the* reason for pulling the plug on the supercollider.

Scientific Pork

A consistent issue that grew more persistent as the hearings and debates reached their and the collider's own denouement was the choice to be made between so-called "big science" and "small science." The late 1980s and early 1990s, when this debate took place, saw government deficits at their highest levels in history. Most legislators believed that voters were tired of big government spending and were not willing to pay higher taxes to discover the origins of the universe. Given a finite fiscal pie, opponents argued, as the supercollider took bigger and bigger bites, less would be available for other programs. Virtually all the legislators who spoke against the SSC voiced support for the project in principle but considered it too exorbitant in fact. Senator Craig of Idaho went so far as to ask whether the "economic value of SSC" was worth sacrificing educational programs such as Head Start, nutritional programs for the poor, or immunization programs for children. Although he insisted that he believed in the SSC's mission, he lamented that fiscal constraints left Congress with "these kinds of choices."[15] The issue thus became a matter of spending priorities.

The focus on priorities allowed opponents the opportunity to avoid criticizing SSC on the merits. The supercollider could be opposed, instead, in the name of children in need of education or immunizations or the thousands of individual scientists toiling away in their small laboratories without money for test tubes. This moral high ground especially appealed to many scientists who testified before congressional committees.

To be sure, in a world of finite resources, the question of where and on what projects money might be spent is an important one. Both scientists and legislators took pains to emphasize that a vote against SSC should not be interpreted as a statement against the role of government in funding basic research. Opposition to a particular project, therefore, should not necessarily be considered opposition to science. Senator Dale Bumpers, for example, took umbrage at Senator Johnston's reference to SSC opponents as members of "the Flat Earth Society." He responded, "No, I am a member of the flat broke society." Setting priorities, of course, is the quintessential responsibility of Congress. What was so fascinating about the SSC debate in Congress was how ardently scientists—especially many who were not particle physicists—joined this debate.

One opponent, Dr. Paul A. Fleury, director of AT&T Bell Laboratories' physical research division, for example, testified at length about how the SSC would hurt "small science." He explained that in 1992 alone, including funding for the supercollider, high-energy physics would receive $1,245 million. This amount, he argued,

> will support approximately 5,000 high-energy physicists, over half of whom are theorists. In contrast, less than $440 million has been requested to support the more than 20,000 condensed matter and materials physicists, who epitomize what has been called here "small science."[16]

Lost in the enthusiasm for small science, however, was the political reality that a dollar (or 10 billion of them) saved from canceling the supercollider was not necessarily a dollar spent for small science. In fact, contrary to the apparent understanding of many of the scientists who testified against the SSC, there was simply no direct connection between the SSC and other science projects. It was not as if Congress had decided to spend $50 billion on science and the only question was how that money should be divided. Senator Johnston highlighted this fact to Dr. Anderson, an opponent of the SSC, who was testifying before him: "In our disorganized system up here, you understand that [the SSC] does not compete with the National Science Foundation and NIH, except in a very indirect sense, coming as it does in a different jurisdictional appropriations subcommittee, so that if we kill this, this money would not revert to one of those, it would revert . . . to [some] water project."[17] Anderson understood this point and fairly maintained that "the actual decision is up to you."[18]

Proponents argued, in contrast, that the supercollider was not being built at the expense of small science. Dr. Allan Bromley, director of the

Office of Science and Technology Policy, and W. Henson Moore, deputy secretary at the Department of Energy, both insisted before Senate committees that small science was not being sacrificed for the supercollider. They pointed out that appropriations for fiscal year 1991, for example, had increased by 50 percent. Moreover, as Moore testified, "No science has been cut or is being cut for fiscal year 1992 to reach the funding levels necessary to get on with this project."[19] Finally, and most important, they argued, "The benefits we are seeing from the Super Collider benefit small science." Moore explained that SSC funding is "directly going to some 90 universities across the country as grants to do research work in support of this project."[20]

Anderson, Fleury, and the other scientists who testified against the supercollider were somewhat naive in the ways of politics. They reached the conclusion that the SSC was "not worth it" because they were proceeding on the assumption that the same amount of money might be better spent on other science projects. But this was never really the issue. To conclude that a particular sum is not worth spending, you have to have some idea of the relative value of the money and the alternatives for the expenditure. This was a perspective that the scientists never had and never could have had. Anderson was correct in saying that the decision had to be for Congress alone. He should have left it at that. But he and other opponents continued to err in permitting their testimony to veer into advocacy of specific spending priorities. The scientists could offer and should have offered no more than to state their opinions regarding the value of the SSC and the need to spend money on science. By offering opinions about comparison shopping, they allowed themselves to be used by those opposing SSC for reasons having little to do with the choice between big science and small science.

In the end, the scientific pork of the supercollider suffered from much the same problem as the political pork: too much money going to too small a group, with those not in on the sharing of the bacon having sufficient political power to ruin the feast. The particle physicists were the Texas of the scientific community in more ways than one. When you consider the many scientists vying for federal dollars, SSC left a lot of researchers with their noses pressed against the window peering in at the billions being spent by a group of scientists already renowned for their elitist attitudes. It did not help matters much that the project directors spent tax dollars lavishly, including the often-cited example of $560,000 for potted plants. The arrogance of particle physicists is generally supposed. They are the loftiest priests of the priestly scientific community. Lee Smolin captures some of this smugness in his description of why, in his youthful hubris, he decided to study particle physics:

When I was trained as an elementary particle theorist, I believed myself to be joining the exalted ranks of those whose task is to discover the fundamental reality behind our perceptions of nature. I always felt a bit sorry for scientists who were not elementary particle theorists, for I could never understand how they could find complete satisfaction in investigating nature at any other than its most fundamental level.[21]

Killing the SSC thus constituted a strike against the conceit of both Texas and particle physics, at the same time standing up for the lone scientists toiling away in their small laboratories.

But the distinction between big and small science is not entirely clear. As used in the supercollider debate, the distinction appeared to refer to two very different issues. Legislators tended to use these terms to describe big-ticket items under a single umbrella, such as SSC, the space station, and the human genome project. All of these multibillion dollar projects lived or died in toto, and how widely dispersed the funds were to small research laboratories was somewhat beside the point. Legislators therefore equated big-ticket costs with big science spending. Scientists, on the other hand, tended to describe big science spending as projects that concentrated the wealth into a few scientists' hands. The supercollider and the space station were therefore big science because the funds did not get spread out sufficiently to many laboratories and scientists. In contrast, the human genome project, which assumed the task of mapping the entire human genetic code, was deemed to be "small" by many scientists who testified against the SSC because funding was spread widely among many scientists.

From a policy standpoint, these different perspectives concerning spending priorities are important if only to make clear what the debate is about. From the congressional viewpoint, there should be no reflexive response against big-ticket science. Some projects, to be done at all, require huge investments. Mapping the genetic code, no less than building the B-1 bomber, is expensive. The legislative question must be "Is it worth it?" This big-ticket question is largely separate from the issue of who receives or how many receive the money. Some projects, like the supercollider and space station, are highly centralized because of the nature of the endeavor. But even with highly centralized projects, discrete research problems were subcontracted out to scores of small laboratories. Ultimately, if politics demands and science allows, most big-ticket projects can be dispersed widely to support small laboratories.

For some scientists, however, the problem with big-ticket science is that it sets research agendas and diverts attention from other projects. Scientists pride themselves on following hypotheses and experiments wherever they lead. Science, in this romantic vision, is a decentralized market in which individual scientists set their own agendas. Congress, as the 800-pound gorilla, upsets this balance by stepping in with its billions and setting forth, much like a central planner, what hypotheses to study and the ways in which to study them. This view of Congress as agenda setter, however, is too simplistic and does not describe the reality of congressional science spending. It should be recalled that it was the particle physicists who first approached Congress with the idea for a supercollider. Legislators have little or no scientific imagination.

Several scientists who opposed the SSC testified that many of the major advances in science are attributable to small laboratory research and that most of the Nobel prizes have rewarded work done on a small scale. Surprisingly, these assertions were never accompanied by citations to research documenting these facts. But even if, on average, it is true that small science produces superior results, this is hardly an argument for not doing big science. While small laboratories do not build super-colliders, launch Hubble telescopes, or explore Mars, they do still benefit from these expenditures. Science operates along a range of practice sizes. As a collaborative effort, it is unfortunate when Congress attempts and scientists permit fields to denigrate one another under the illusion of competing for precious resources.

The big-science/small-science issue was really a debate among scientists that lawmakers interpreted incorrectly. Scientists who testified against the SSC never fully appreciated this legislative error. For Congress, big science was expensive science, which made it bad policy. Scientists, however, do not think of small science as necessarily being less expensive than big science. Legislative opponents of the SSC were thus driven by the thought that "we can kill the SSC and help balance the budget with the savings." Scientist opponents were driven by the thought that "we can kill the SSC and support many scientists with the savings." The only thought they had in common was killing the SSC. And that is what happened.

The fact of the matter is that the supercollider was huge and could not be done at modest cost. The same is true of the space station. Hence, the policy issue was whether the SSC was worth building. In its less than infinite wisdom, Congress decided it was not. As a science enthusiast, I personally agree with Leon Lederman, a Nobel laureate and emeritus director of the Fermilab, who commented, "It's disheartening that a large number of fairly intelligent people could do such a dumb thing."[22] The supercollider was an exciting project that, like all similarly ambitious endeavors, offered potentially profound benefits at high risk. But if they had to decide against SSC, as a taxpayer I wish they could have done the dumb thing $2 billion earlier.

Lost in Space

In October 1993, Congress embarked on a new era of space exploration with its decision to authorize funding so that NASA could begin construction of a space station. The space station will be the platform from which most major expeditions to explore our solar system will begin. The space station, however, is not exclusively an American adventure. Japan, Canada, Russia, and the European Space Agency will contribute money, material, and personnel to this grand project. It is thus truly an international, human adventure—and therein lies part of the problem. Ironically, given the imagination necessary even to contemplate space exploration, this area of legislative science suffers from what former President Bush referred to as "that vision thing."

The space station has had two sets of critics. Like the Christian right joining with radical feminists to oppose pornography, these critics of the space station make strange bedfellows. One group, which numbers many congressional representatives among it, is not particularly fond of spending money on scientific research period, and the space station for them represents a classic example of federal largesse. The other group, which includes some legislators and many scientists, strongly supports expansive federal funding of space exploration but considers the costs of including human explorers to be needlessly wasteful.

Every year since the original funding was approved, a group of representatives led by Senator Dale Bumpers (R., Arkansas) has tried to terminate funding for the space station. They complain about the astronomical costs involved (which they quote as likely to approach $100 billion), the unreliability of our partners (especially Russia), and the need to concentrate on earthbound needs (usually left unstated beyond general sentiments about balancing the budget). These critics discount the value of past inventions derived from space travel and doubt the benefits of escaping the confines of earth.

Critics who support substantial federal spending on space exploration but question the need to put humans in space argue that more could be done by relying on unmanned space travel. The human element in space exploration, according to this view, adds tremendous costs without concomitant benefits in terms of information gained or discoveries found. Many eminent scientists subscribe to this view. For example, the American Physical Society, to which virtually all U.S. physicists belong, issued an official statement in January 1991 opposing NASA's reliance on manned space exploration. They urged that "the potential contributions of a manned space station to the physical sciences have been greatly overstated" and concluded that earth-based and robotic-reliant research would accomplish the task as well or better. Nobel laureate Steven Weinberg noted that "if from the beginning it had been planned that an unmanned rocket instead of the space shuttle would put the Hubble Space Telescope into orbit, then seven similar telescopes could have been built and launched for what so far has been spent on just one."[23] Similarly, Robert Park, a physicist at the University of Maryland, complained that the public fascination with manned space travel is a product of the "star trek myth."[24] Commenting on mechanical exploration of Mars, he pointed out that humans in space are capable of no more than robots and probably less. And robots can do it safer and cheaper than humans.

These two sets of critics, however, ignore certain basic truths of space exploration. The first lacks an appreciation for the great, albeit unknown potential of space travel. The second lacks an appreciation of the allure of space travel for the average person. Together, these critics possess a decidedly unromantic conception of space exploration.

Legislative critics who repeatedly attack the costs of the space station have fairly pedestrian views regarding the benefits of space. They measure economic efficiency simply by the benefits they can anticipate or can demonstrably attribute to past space travel. These legislators would have fit in well with the treasurers in Queen Isabella's court who, in 1491, decried the costs of supplying Columbus's expedition. Indeed, Columbus—who sought a new route to Asia but instead found a "new world"—is a poignant example of how the treasures of exploration can never fully be anticipated.

If legislative decisions to invest in exploration were limited to those projects for which rewards could be readily anticipated, we would live in a very different United States. John Glenn, the first man to orbit the earth and now the oldest man to go into space, and a Senator from Ohio, illustrated by example the sort of shortsightedness demonstrated by proposals to cut off space station funding. He quoted Daniel Webster's opposition to federal funding of land acquisitions West of the Mississippi:

> What do we want with this vast worthless area, this region of savages and wild beasts, of deserts of shifting sands and whirlwinds of dust and cactus and prairie dogs? To what use could we ever hope to put these great deserts or those endless mountain ranges, impenetrable and covered to their very base with eternal snow? What can we ever hope to do with the western coast, a coast of 3,000 miles, rock-bound, cheerless, uninviting, and not a harbor on it? What use have we for this country? Mr. President, I will never vote 1 cent from the Public Treasury to place the Pacific coast 1 inch nearer to Boston than it is now.[25]

Critics who lament the amounts spent on manned space exploration, mostly well-meaning scientists who measure success by the amount of information gleaned for each dollar spent, do not understand the romance space travel has for the general public. It is undoubtedly true that most cosmologists, astronomers, and physicists are romantic in their quest for knowledge about the origins and structure of our universe. But for many scientists, the romance is in the findings, not the adventure of the search. Many nonscientists share this desire for new information, but manned space exploration holds a special place in the public imagination. John Glenn, responding to the Webster/Bumpers' myopia, argued earnestly,

> People have stood here on Earth and looked up for a hundred years, or several hundred thousand years. We have wanted to travel up there. We wanted to go see what it was like. Now we can use that area of space.[26]

Although John Glenn apparently includes *Homo erectus* or Neanderthal man in his romantic vision, since modern man evolved only within the last 120,000 years, his point is well taken. Indeed, though probably not intended, Glenn's statement conjures up images of the famous opening scene in *2001: A Space Odyssey*. In that scene an apelike man (probably Australopithecus africanus) finds a large bone lying near a pile of other bones. He picks up the bone and, realizing its lethal potential, begins smashing the other bones to smithereens, all to the accompaniment of *Thus Spoke Zarathustra* by Richard Strauss. In the next scene, a band of these bone-wielding creatures is shown driving off an unarmed group and, in the process, beating to death one of these evolutionarily hapless animals. One of the evolutionarily fortunate apes, in celebration, throws his bone into the air. The camera follows its ascent in slow motion. As it reaches the apex of its flight, the bone transforms into a spaceship, which is then followed by a ballet of dozens of spaceships and space stations. One of these spaceships, of course, contains Hal, a very humanlike computer, and a human crew. This last point should not be overlooked. Although Hal was in some ways the star of the movie, *2001*, like every other major science fiction film that followed it, includes human explorers.

It may be true that one space station would pay for a hundred Hubble telescopes. It is not obvious, however, that the boon in information would excite the average taxpayer's imagination as strongly as, say, Neil Armstrong's stirring words as he stepped onto the moon: "That's one small step for [a] man, one giant leap for mankind." Unexcited taxpayers do not agree to purchase a hundred Hubble telescopes, however cost-effective they might be. Similarly, politicians who would prefer to remain pedestrian earth-dwellers and save-a-buck too misunderstand this romantic zeal. As Oscar Wilde put it: "We are all in the gutter, but some of us are looking at the stars."[27] Our history as a species is one of exploration and migration. This restlessness has not always led to salutary results, perhaps, but it seems an essential part of our character. In federal policy making, this character will almost certainly drive us to explore the heavens. This exploration, however, will never be the most cost-efficient in terms of information gained for dollar spent. Frail, untidy, and inefficient humans still need to go along for the ride.

Cloning; or, The Modern Prometheus: The Legislature Setting the Moral Boundaries of Science

Did I request thee, Maker, from my clay
To mould me man? Did I solicit thee
From darkness to promote me?

—JOHN MILTON, *Paradise Lost*

While the legislature often indirectly dictates a scientific agenda through its spending decisions on projects like the supercollider and space station, sometimes it ventures directly into the fray and considers legislating the moral boundaries of scientific research. Few issues of recent times have excited public reaction greater than the news that researchers in Scotland had successfully cloned an adult sheep. The offspring was named Dolly and, as the headlines blared HELLO DOLLY, it was clear that a new era was upon us. Or at least it seemed clear. From the headlines, a reader might have reasonably concluded that somewhere in the Scottish highlands, Dr. Ian "V. Frankenstein" Wilmut had already "beheld the wretch—the miserable monster whom [he] had created."[28] Although cloning humans remains technologically many years off, legislators and the president wasted no time projecting the worst.

Before news of Dolly had time to travel once around the Washington Beltway, President Clinton asked the National Bioethics Advisory Commission to evaluate cloning. Committees in both houses of Congress held hearings on the use of cloning technology to create human beings, and several bills were introduced ranging from prohibitions on funding for human cloning research to outright federal criminalization of such research. The leaders in Washington had heard the dire warning of Dr. Frankenstein:

You seek for knowledge and wisdom as I once did; and I ardently hope that the gratification of your wishes may not be a serpent to sting you, as mine has been.[29]

Congressional responses to cloning have varied, but the greatest sentiment clearly is against the pursuit of human cloning. Even Dr. Wilmut, in testimony before the Senate Committee on Labor and Human Resources, concluded summarily that "cloning of humans would be unethical."[30] Many legislators have joined this conclusory chorus, finding the issue to be sufficiently obvious that it should be foreclosed without further adieu. Senator Bond, for instance, exclaimed firmly, "I do not think we need to study this." He then introduced legislation to prohibit research on cloning humans altogether. He explained:

> I intend to make sure that human cloning stays within the realm of science fiction; and does not become a reality, regardless of whether or not it is technically feasible to clone a human being. . . . This type of research on humans is reprehensible, and we should not be creating human beings for spare parts, as replacements, or for other unnatural and selfish purposes.[31]

Cloning provoked a number of appeals to religion among the legislators, as well as general consternation over man playing God. Congressman Ehlers, for instance, observed that "any discovery that touches upon human creation is not simply a matter of scientific inquiry. It is a matter of morality and spirituality as well."[32] Senator Bond emphatically echoed this sentiment, declaring, "Humans are not God and we should therefore not try and play God."[33]

The moral ramifications of cloning are an important consideration for Congress, since the technical ability to accomplish something has never been a very good argument for doing it. Somewhat troubling, however, is the lack of interest among some legislators in the full scientific details of human cloning. Legislators leaped to oppose the idea before it was ever explained. It seems that in trying to avoid science fiction outcomes, legislators have allowed these nightmares to guide their thinking. We need not see virtue in Dr. Frankenstein's "miserable wretch" to believe that human cloning might have enough potential that it is worth further study.

An often-repeated challenge to cloning concerns the welfare of the children that would result from the process. Dr. Harold Varmus, Director of the National Institutes of Health, made this argument when he maintained that the "notion of carrying out cloning human populations, to my mind, is not consistent with the traditional ideas of human individuality and diversity."

But the power of an argument that is based on the cloned child's best interests is not obvious. This is especially the case given that the alternative for that child is nonexistence. Clones would simply be identical twins of the "parent." Only someone who believed that we are completely a function of the sum total of our genes could suggest that a clone was predestined to some single fate. Although research on twins indicates that genes explain a large percentage of human behavior, experience remains an overwhelming force in the development of human personalities. Certainly identical twins are not inherently objectionable. Children conceived through the technology of cloning would be autonomous individuals, allowed to define their own futures.

The more important issue concerns equal treatment of these children. Just as there is no basis for treating "test-tube babies" differently, "cloned children" should similarly be treated in a fair and equal fashion. They would be human beings with all the rights and liabilities enjoyed by other individuals. Cloning, like test-tube conception before it and artificial insemination before that, engenders visceral reactions largely out of proportion to the realities of the technology. Pope Pius XII, for example, denounced artificial insemination even between husband and wife: "To reduce cohabitation of married persons and the conjugal act to a mere organic function for the transmission of the germ of life would be to convert the domestic hearth, sanctuary of the family, into nothing more than a biological laboratory."[34] Pope John Paul II similarly condemned human cloning, stating that it would be an affront to the dignity of man.

A call for further study and sober reflection does not mean that cloning necessarily is a good idea or that some regulations might not be needed to control this new technology. It is the rush to judgment that is objectionable. An essential concern with cloning is that some people might wish to employ this technology for distasteful reasons. It is not terribly difficult to imagine certain personalities who are so taken with themselves that they would consider it a gift to the world to clone themselves. Immortality is an alluring object, and cloning might be perceived as an avenue to achieve it. Such abuses are foreseeable and probably not easily controlled. But the evil lies in us, not the technology.

Cloning raises few new ethical issues apart from what other technologies present, and it might even resolve some. Consider, for instance, the celebrated case of the Ayala family. In 1989, Abe and Mary Ayala learned that their sixteen-year-old daughter, Anissa, suffered from chronic myelogenous leukemia, an illness that kills 90 percent of its victims within five years. The only known therapy was a bone marrow transplant. However, neither the Ayala parents nor their eighteen-year-old son were suitable matches, and a nationwide search turned up no donors.

The Ayalas decided to conceive a third child for the express purpose of having a child that would be a match and give Anissa a chance. The probabilities were 1 in 4 that the child would match. Doctors who specialize in doing marrow transplants report that this is not a rare practice. One study found over forty cases in a five-year period in which parents conceived children for the purpose of bone marrow transplantation. The Ayalas were unusual in that they announced their intentions and so subjected themselves to public scrutiny. They were also unusual because Mary was forty-two years old and Abe had had a vasectomy. These facts lengthened their odds of conception considerably. Doctors were able to reverse the vasectomy and Mary and Abe were successful in conceiving their third child within a relatively short period of time.

In April 1990, the Ayalas gave birth to Marissa, a healthy baby girl, who was a suitable bone marrow donor for her sister Anissa. At fourteen months, Marissa underwent an uncomfortable but quite safe procedure to draw bone marrow from her hip to be donated to her ailing sister. The procedure was a success and Anissa's leukemia went into complete remission. The Ayalas acted much the way we would expect most parents would have acted. They tried to use the best technology science could offer to save their child. Anissa explained it best: "My parents did the only thing they could do. I look on them as heroes." If cloning had been available and there were no other viable alternatives, the Ayalas almost certainly would have used it to try and save their daughter's life. If we can sympathize with such a judgment, can we reflexively reject the technology that would give parents the opportunity to make this decision?

The principle objection to the Ayalas' decision to have a donor child is the same as the principle objection to cloning, treating the child as a means rather than as an end. Reacting to the Ayalas' action, for instance, the Reverend John Fletcher, director of the Center for Biomedical Ethics at the University of Virginia, argued:"This represents the 'thingification' of a child. . . . It assaults some very deeply held values." But this perspective reflects a rather crabbed understanding of parenthood and family values. People have children for a variety of reasons, many substantially less admirable than the Ayalas'. If it is acceptable to have a child to work the fields or for security in one's old age, is it wrong to have a child to save another child's life?

In the context of the Ayalas' situation, cloning still presents the ethical quandary of having a child to serve another's interest. It would, however, avoid the attendant ethical dilemma whether to abort a nonmatching fetus. For while the Ayalas stated that they never considered aborting the fetus if it had not matched, their decision is contrary to many other parents in this situation who legally decide to abort nonmatching fetuses until a match is produced. Indeed, in one case, parents who had a child for the express purpose of saving an older sibling subsequently put the donor child up for adoption. Cloning, in contrast, would guarantee the match and eliminate many ethical dilemmas. It also would overcome probable barriers such as if Mary had been too old to have another child or Abe's vasectomy could not have been reversed. Cloning would not create the moral quandary of treating children as means rather than ends. Nor would cloning introduce reproduction through technology. In fact, virtually every significant moral obstacle raised against cloning is present with technologies now routinely employed. Those who would ban cloning completely have the burden of explaining why it should be singled out for such condemnation.

It was not surprising, perhaps, that the public's initial reaction to cloning, a truly radical advance in biotechnology, was formed by popular fiction. Cloning has been so long the stuff of science fiction and, indeed, not seriously considered a subject of science at all that Wilmut's accomplishment invoked a reflexive response. But the reality of this technology is more complicated and more subtle than popular fiction captures. Cloning, as does all new technology, offers both promise and peril. The twentieth century witnessed technical revolutions ranging from splitting the atom to splitting the DNA molecule, all of which present equal parts heaven and hell. Cloning is just one more technical revolution. It is not that "miserable wretch" Dolly we should fear. It is our own ignorance that should make us tremble.

Candy Is Dandy[35]: Science as a Component of Legislative Policy Making

In the winter of 1977, fourteen Canadian rats were diagnosed with bladder cancer. The afflicted animals were part of an experimental group of ninety-four rats that had been fed a diet containing large doses of saccharin. In fact, saccharin made up 5 percent of their diet. None of the eighty-nine rats that had been fed a diet free of saccharin developed similar cancers.[36] Largely in response to these findings, the FDA announced on March 9, 1977, a proposed ban on saccharin. The response from Congress and the nation generally was immediate. The public was not happy.

The public made it abundantly clear that it had no intention of giving up its diet Coke without a fight. In a nation obsessed with both weight and cancer it was fascinating and a little disturbing to see which malady the public feared more. Still, there was much about the Canadian study that failed to inspire confidence. The FDA itself, in an effort to avoid any panic, explained that the rats had been fed the equivalent of 800 cans of diet soda a day. Rather than tempering fears, however, this statement led to derision of both the study and the agency that based such a momentous ruling on such nonsense. Editorial cartoonists depicted bloated rats staggering around holding cans of diet soda. One legislator suggested a label warning "The Canadians have determined saccharin is dangerous to your rat's health." Twelve days after the announcement, Congress began holding hearings to consider voiding any proposed ban and possibly amending the law on which the ban was based.

At the start, the FDA adopted the position that it really had no discretion in the matter. The fault lay with the law the agency was charged with administering and thus, implicitly, with Congress itself. Under the Food, Drug and Cosmetic Act, the safety of food additives must be established by the manufacturer and, as provided in the 1958 Delaney Amendment, "no additive shall be deemed to be safe if it is found to induce cancer when ingested by man or animal, or if it is found, after tests which are appropriate for the evaluation of the safety of food additives, to induce cancer in man or animals." This had the effect of focusing much of the initial congressional effort on modifying the Delaney Amendment. Soon, however, the FDA made clear that even absent the Delaney language, the agency was probably compelled to ban saccharin under the general provisions of the law. According to the research, saccharin rendered food unsafe. The debate thus moved to the merits of the science and more particularly to the merits of the proposed ban.

In addition to the huge doses of saccharin the rats ingested, critics questioned the value of the science on a number of other grounds. For example, legislators questioned the propriety of generalizing from rats to humans. The evidence from human epidemiological studies was scant, and in fact several human studies of diabetics suggested a lower incidence of bladder cancer among saccharin users. Finally, the FDA could not articulate any threshold level of saccharin use that might be safe. The research simply did not provide an answer to the persistent question among legislators: What about people who drink less than 800 cans of diet soda a day?

Although there was no smoking gun, critics' attacks against the science were somewhat overheated. Foremost, while the discussion always seemed to return to the hapless Canadian rats, at least six other studies, dating to the early 1950s, indicated a link between saccharin and cancer. Some of these were the FDA's own studies, which originally had been doubted due to anomalies in the research methods. The Canadian study became the focus of public attention because, for the FDA, it was the final nail in saccharin's coffin. But, as is true generally in the world of applied science, no one study "proves" any hypothesis. The real question is whether the weight of authority has reached sufficient critical mass to support the decision that must be made. For this reason, congressional critics' attacks on the FDA's reliance on rat studies was entirely misplaced. Animal studies are an integral and essential aspect of toxicological testing. Again, the issue was whether these studies were *enough*, not whether they were relevant.

But this criticism also extends to the FDA. When the public erupted in righteous indignation at the thought of a saccharin ban, the FDA sought refuge behind the letter of the law. Sherwin Gardner, the acting commissioner of the FDA, testified before a House subcommittee that the law gave the agency no choice, since the studies "demonstrated beyond serious question that saccharin causes malignant bladder tumors in test animals."[37] And the law "provides unequivocally that a substance which has been shown by appropriate tests to cause cancer in animals may not be permitted in foods."[38] But this conclusion reflects an overly generous interpretation of the research, too strict a reading of the law, and a modest view of the FDA's authority over food additives. The pertinent words of the law require the FDA to ban additives that, "after tests which are appropriate," are found "to induce cancer in man or animals." In deciding which tests "are appropriate," the FDA retains a broad measure of discretion. This power was evident in the FDA's failure to ban saccharin based on the prior studies.

The real problem in the saccharin wars of the 1970s was that neither the FDA nor any other institution had clear responsibility for asking the most pertinent question in light of the research results: Given the risks posed by saccharine and the benefits derived from its use, is it a good idea to ban it? The FDA itself had no power under the law to do a cost-benefit analysis of any ban. As Richard Merrill, the FDA's general counsel, remarked, "We ask ourselves only: 'Is it shown to be safe for human beings?' And we would evaluate saccharin—we are forced to evaluate saccharin under that standard in the same way we would be evaluating the seventeenth emulsifier for fortified bread."[39] Since the FDA was proscribed from conducting the cost-benefit analysis, this task fell, by design or default, to Congress.

Spurred by the public's outrage, Congress assumed the task of evaluating the costs and benefits of saccharin. And, indeed, Congress on the whole did a pretty good job of sifting through the evidence, weighing all points of view, and carving out a reasonable compromise. Congress's success in mediating the saccharin hostilities offers valuable lessons for wars fought on other science policy battlefields.

To begin with, the legislators did what politicians do best: they wetted their index fingers and raised them in the air to see which direction the political winds were blowing. They did not have to wait long for the wind to reach hurricane force, mostly blowing against the FDA. The congressional hearings on saccharin are filled with quotations from letters sent by constituents who opposed the ban. These letters primarily cited two substantial costs of the proposed ban. Of greatest political weight, many diabetics and parents of diabetics wrote to describe the severe imposition that would be caused by the FDA's action. For instance, the following letter was one of many quoted by Senator Hayakawa (Hawaii) in his comments opposing the proposed ban:

"We have two diabetic boys. They have few luxuries with regard to the food they eat. Meals are the high point in a diabetic's day, and we try to make it as pleasant and as enjoyable for them as we can. So when we top a meal off with a sugar-free drink, it is a real treat to them. They feel just like those of us who are not diabetic and they forget for a few minutes that they have a disease for which there is no cure."[40]

Of somewhat lesser political import but affecting many more voters, legislators cited the impact a saccharin ban would have on a variety of health concerns. It was pointed out by many witnesses, for example, that replacing saccharin with sugar would contribute to substantial tooth decay. In addition, an overriding concern registered by many involved dietary issues, especially issues associated with overweight. Often, in order to buttress the argument, high blood pressure and heart disease were thrown into the mix, since these ailments are generally associated with obesity. In fact, however, although intuitively appealing, there is no evidence to indicate that saccharin actually assists people to lose weight. Indeed, supporters of the FDA ban cited research that suggested saccharin might actually contribute to weight gain by lowering blood sugar and stimulating the hunger drive.[41] Nonetheless, Representative Andrew Jacobs, Jr. (Indiana), voiced the concern of many legislators when he remarked:

To allow the ban to stand, would be to bang down one more nail on freedom's coffin—and quite likely a few on regular coffins, too. For, in terms of body weight, without the freedom to choose saccharin, many people would be left only with the freedom to be fat and to suffer heart attacks or diabetic deterioration.[42]

The bottom line presented by the proposed saccharin ban concerned a basic value choice. Given that there were no other alternatives, since cyclamates had been removed from the market seven years earlier, a ban on saccharin would mean either sugar or nothing. Nothing was an unpopular option. Although saccharin provided no physical health benefits, the real psychological benefits made an outright ban unpalatable. And if there is one fundamental rule of politics, it is to avoid making unpopular decisions.

Of course, permitting the continued sale of a cancer-causing agent might itself become unpopular in time. Congress could neither be nor appear to be insensitive to the health risks associated with saccharin. In order to assist itself in understanding the details of the research, Congress asked the Office of Technology Assessment (OTA) to prepare a report on saccharin and the regular protocol for testing food additives.[43] This report eventually provided the well of knowledge from which legislators consistently drew information for crafting a measured response to the FDA's proposed ban. Indeed, the usefulness of the OTA's report on saccharin illustrates well the folly of Congress's subsequent action of killing the OTA in 1995 to save a few dollars. The OTA provided Congress with its own evaluation of the scientific facts on which policy had to be made, thus freeing it from reliance on the scientific assessments of agencies or lobbyists. In a system of government based on checks and balances, such an independent source of information would seem to be imperative.

Throughout the congressional and public debates about saccharin, cigarettes were constantly alluded to as a known carcinogen that remained on the market. Senator Richard Schweiker (Pennsylvania) made this point in observing that "many Americans fail to see why saccharin should be singled out for such drastic action."[44] He continued, "Cigarettes, known to endanger human health in many ways, are allowed to be sold freely and openly. Anyone may choose to smoke or not to smoke; all that is required is that he be warned of the hazards."[45] And, as it turned out, cigarettes became the model by which Congress solved the saccharin issue.

With remarkable expediency, Congress passed in November 1977 a law that both prohibited the FDA from banning saccharin based on the Canadian or other then-existing studies and required the following label to be affixed to all products containing saccharin: "Use of this product may be hazardous to your health. This product contains saccharin which has been determined to cause cancer in laboratory animals."[46]

The advantage to Congress of having an independent scientific assessment from the OTA cannot be overstated. It gave legislators detailed background information free of any suggestion of special interest or other taint. Although there is no reason to believe that the FDA had a specific agenda behind their scientific assessment, certainly once they proposed a ban they had the bureaucratic need to defend it fully to outsiders. The OTA panel could offer a fresh evaluation untainted by this bureaucratic circling of wagons. In addition, especially given the prestige associated with the members of the OTA panel on saccharin, other experts from the FDA or industry would be more inclined to remain circumspect about what conclusions they were willing to draw. In short, with the OTA in the background, any battle of the experts that erupted would remain civilized and reasonably focused on true scientific disagreement. This process allowed consensus to emerge and averted much of the excessive hyperbole often associated with congressional debates.

The scientists from the OTA and most of the other scientists and administrators who testified agreed that saccharin was a "weak" carcinogen. In real terms, this meant that the animal studies indicated that about 2 to 3 percent of the 30,000 new cases of bladder cancer each year could be due to saccharin. Such a weak association would explain the difficulty of identifying the association in epidemiological studies. In addition, virtually all the scientists agreed that there were no hard data about the benefits of saccharin. While there was considerable anecdotal support for the value of a sugar substitute, no research had been done to allow an informed conclusion about the consequences of a saccharin ban on diabetes, dieters, or dental health. Finally, witnesses agreed that social scientists had yet to study the effects of warning labels or what type might be effective, especially as regards young children.

What Is to Be Done?

Congress's resolution of the saccharin controversy provides several concrete lessons concerning legislative use of science. Foremost, Congress needs to be well informed. The OTA's report focused the debate and generated considerable consensus among lawmakers concerning an appropriate solution. Although without the OTA Congress would have heard from many scientists, this testimony would have been more partisan than the testimony the OTA provided, and it probably would have been more extreme without the OTA's presence.

The Delaney Amendment was not the source of the problem in this case, though its presence illustrates a source of concern for the lawmaking process where science is involved. Although the social costs and benefits of any piece of legislation are not easily quantified, and some might defy quantification entirely, we can still hope that, at some level, particular legislation creates more benefits than it costs. Indeed, however unhelpful, this might be the definition of a "good" law. But, as the saccharin example illustrates, a significant question raised in our system, especially when complex science is involved, concerns which branch of government should conduct this cost-benefit calculation.

The lawmaking functions between Congress and regulatory agencies are not easily separated where science is concerned. Hence, any conclusions I offer here are tentative until I visit the matter of administrative agencies more fully. Nonetheless, certain conclusions can be offered regarding Congress's dealings with science and scientists that are worth noting.

Congress is first and foremost a political body. And the realities of politics in legislative bodies are not likely to change in the foreseeable future. Scientific research and understanding, though central to so much of the legislative task, are likely to remain on the periphery of the average legislator's agenda. The business of politics is rhetoric. Science plays at best a supporting role in the political drama. Legislators make decisions and appeal to the public through a discourse meant to persuade the public to accept some conclusion, not convince the public of the accuracy of the basis for that conclusion. So long as the general public remains persuaded without a fulsome exploration of the details of the science, legislators will feel little pressure themselves to come to grips with the details. Legislators are therefore likely to continue to deal with science in a halting and slapdash manner. Their weapon of choice, and probably the only weapon they will ever have available, is a blunt sword.

The fact of the matter is that democracy is a sloppy and inefficient way to manage a society. H. L. Mencken observed, "Democracy is the theory that the common people know what they want, and deserve to get it good and hard."[47] The swirl and tangle of democratic politics ensures that the sometimes priestly statements from the scientific temple will be ignored. At the same time, science is never entirely free of politics either, and must be understood as merely one tool among many that serve a democratic polity. Science is replete with value judgments ranging from the hypotheses studied to the standards applied for concluding that an "answer" has been "discovered." Few political scenarios are quite as disturbing as a government run by a scientific oligarchy. However chaotic the system might sometimes seem, popularly elected legislators must provide essential oversight to the implementation of scientific policy. Winston Churchill, a leader well versed in the weaknesses of democracy but all too aware of the alternatives, summed up this insight as follows:

No one pretends that democracy is perfect or all-wise. Indeed, it has been said that democracy is the worst form of Government except all those other forms that have been tried from time to time.[48]

That Congress deals with science with a blunt sword, then, is not altogether bad. Its membership is largely trained in the art of legal rhetoric, with little or no aptitude or interest in the niceties of the scientific method. And institutionally, Congress, so long as it remains democratic, is incapable of responding to science with the level of complex understanding that the subject requires. Congress, therefore, should and largely does play an interested bystander role in relation to science policy. Bystander because it does not know enough to participate fully. Interested because it is duty bound to oversee the development of science policy and from time to time must step into the fray to correct perceived anomalies, defects, or errors. The real instrument of science policy in the United States is the bureaucracy of the executive branch of government.

Administrative agencies do the heavy lifting of scientific policy formation and implementation. Congress tends to set broad objectives and leaves to agency rule making the rather more difficult job of marshaling the science and managing the interest groups in order to achieve these goals. But Congress's role does not end with this delegation of authority. Congressional staff and committees and sometimes Congress itself regularly provide oversight to this rule-making function of the executive branch. In this capacity, Congress often bludgeons bureaucrats and scientists alike, sometimes for very good reasons. Most of the time, however, it swings wildly but without causing much harm.

VI

RARELY PURE, NEVER SIMPLE
Science in the Federal Bureaucracy

I guess this is why I hate governments, all governments. It is always the rule, the fine print, carried out by fine-print men. There's nothing to fight, no wall to hammer with frustrated fists.

—JOHN STEINBECK

The actual work of government is too unglamorous for the people who govern us to do. Important elected officeholders and high appointed officials create bureaucratic departments to perform the humdrum tasks of national supervision.

—P.J. O'ROURKE

The regulatory state we know so well today, administered by tens of thousands of federal bureaucrats, is almost entirely a creature of the twentieth century. The American founders did not provide for it. Even *The Federalist* does not anticipate its existence, despite its otherwise extraordinary dissertation on the American form of government. In fact, other than the army and the navy, the Constitution, including all twenty-seven amendments added since 1789, is virtually silent regarding the federal bureaucracy. The modern administrative agency developed largely in response to the increased centralization of government and the increasing technological challenges posed by the twentieth century. In fact, agencies were largely created to deal with the technical details and complex technological and scientific aspects associated with the ever-expanding federal juggernaut, especially following 1932 and Franklin Roosevelt's "New Deal." They were staffed with experts who could understand the complexities necessary to the day-to-day implementation of laws that Congress did not have the expertise or institutional competence to handle. In short, agencies emerged largely to manage the science and technology that was beyond the abilities of the Congress, the president and the judiciary.

Administrative agencies today control an overwhelming proportion of the governmental details of the average American's daily life, ranging from the United States Postal Service (USPS) and the Internal Revenue Service (IRS) to the National Highway Traffic Safety Administration (NHTSA). In addition, they control much of the policy-setting agenda regarding some of the most profound aspects of America's future, ranging from the Equal Employment Opportunity Commission (EEOC) and the Environmental Protection Agency (EPA) to the National Institutes of Health (NIH). Administrative agencies are, for all intents and purposes, a fourth branch of the national government. They receive — are delegated — their authority from Congress, but they are situated within the executive branch under the president's supervision. Consistent with the checks and balances inherent in all parts of the federal government, agencies are subject to the direction of the president, the oversight of Congress and, to a lesser extent, review by the courts.

Although agencies have much in common (beyond their obsession with acronyms), they exhibit a wide variance in their actual practices.[1] Thus, lessons about one agency's scientific prowess or lack thereof will not always be generalizeable to other agencies. Nonetheless, there are valuable lessons to be learned by comparing agency use of science to legislative and judicial use of this information. While agencies are hardly temples of sophisticated empiricism, they far outperform the dens of empirical iniquity sometimes found in the judiciary and that pervade the Congress.

The Center of the Storm of Government

The key to understanding federal agency practice is to see where agencies are situated in the federal government. Agencies are created by Congress and are given mandates to carry out certain objectives. These mandates are sometimes quite specific, such as the National Railroad Passenger Corporation, the federal umbrella for Amtrak, whose only mission is to manage "America's national passenger railroad." Typically, however, Congress invests agencies with a very broad mandate, such as the EPA's mission to "protect public health and to safeguard and improve the natural environment." Since agencies are empowered to implement the law, they are technically within the executive branch of government and thus under the president's direct supervision. In practice, however, both the president and Congress play strong supervisory roles. The president is the immediate boss of most agency heads but typically cannot remove those administrators arbitrarily or even for political reasons alone. Congress has strong supervisory powers over agencies, since it controls the grants of authority and budgets that enable agencies to operate. Since Congress, however, is composed of 100 senatorial managers and 435 representative managers, it tends to lumber about awkwardly and bluster incessantly about agency practice, usually without substantive effect on most agency decisions.

Congress typically directs agencies to solve some problem, oversee some business practice, or monitor some situation. It usually does no more than suggest the broad outlines, since it does not have the time, expertise, or, too often, political will to give detailed directions. Thus, the FDA is empowered to oversee clinical drug trials and make safety assessments, the FTC to provide detailed reviews of proposed mergers for compliance with the antitrust laws, and the National Fish and Wildlife Service to evaluate whether certain species are endangered and thus qualified for protection under the Endangered Species Act. Pursuant to these various mandates, agencies promulgate a torrent of federal regulations (codified in the Code of Federal Regulations, or CFR). These regulations, in turn, are a product of an elaborate dance of public comment and lobbying efforts that are mandated by a generally applicable law called the Administrative Procedures Act (APA). Although many agencies have their own procedures, the APA controls most of the rule making they do. Moreover, although not central to the story told here, many agencies are also empowered to prosecute and try alleged wrongdoers. Therefore, for example, EPA promulgates rules against polluting, decides who to prosecute for breaking the rules, and provides the trial process by which violators are punished. Although all three responsibilities receive some degree of oversight by Congress, the president or the courts, agencies have awesome power in being lawmaker, prosecutor, and judge in their respective domains.

On the lawmaking side, the picture that emerges of agencies is of rule makers who are either themselves technically proficient in their areas of governance or have large professional staffs or scientific advisory committees who provide them with this technical detail. These technocrats are not elected and are subject to only indirect political accountability. Yet albeit indirect, this political accountability can amount to a barrage of cannon fire shot from a multitude of directions, including Congress, the president, affected businesses, interested citizen groups, and virtually any person with a fax machine, e-mail address, or first-class postage stamp. Surprisingly, then, from this chaos comes a substantial amount of quality rule making that is based on fairly good scientific research and that is tailored to the political desires of the electorate.

Too Many Cooks Spoil the Broth

The many power centers of the American constitutional system are designed to check and balance one another. Indeed, it is perhaps the central and most profound insight of the drafters of the Constitution that the best way to control government is to divide it into parts that will compete for the power and the affection of the people. We divide power between the federal government and the states; and within the federal government power is divided between the president, the Congress and the courts. But power is divided still further. States also have their divisions between executive, legislative, and judicial branches. At the national level, the president divides power between himself and the agencies, Congress is divided into two houses and further divided into a myriad of committees and subcommittees, and the courts are separated into trial courts, intermediate appellate courts, and the Supreme Court. It is truly a wonder that government ever gets anything done.

But with these many power centers, sometimes too much gets done, or at least everyone is working on the same problem at the same time. This usually happens when the matter is not particularly controversial but is sufficiently outrageous that everyone wants to take credit for putting a stop to it. Some examples include child pornography, police misconduct, illegal drugs, tobacco, and, for a time, paparazzi mistreatment of celebrities. Cloning, considered in Chapter V, is another example. But not all subjects, no matter how egregious, lend themselves to the mechanisms available to the respective departments. For example, it might be that testing school children regularly in math and science is a good idea to ensure educational standards; but it is not clear that these standards should be imposed by the national government in light of our tradition favoring local control of education. Similar questions about locus of control over lawmaking arise constantly between the states and the national government, with examples such as gun control, illegal drugs, domestic violence, food labeling, immigration, waste disposal, highway speed limits, and abortion constantly being debated and regulated. There are similar locus of control questions between the executive and legislative departments of the federal government. High-profile problems often garner the attention of Congress and one or more federal agencies. Some problems are especially well suited to the blunt congressional sword, while others would be bludgeoned by such attention and need instead the scalpel of administrative rule making.

Returning to cloning to illustrate this point, let us assume that it is a good idea to prohibit the technique's application to humans. If cloning humans is to be prohibited, a substantial question arises over which department would be best at defining the ban. Congress, as we saw previously, debated bills that would have prohibited the use of federal funds for cloning or would have prohibited cloning all together. The president, not waiting for Congress or his own federal agencies to act, imposed an immediate ban on the use of federal funds for human cloning research, pending completion of a review of the ethical implications of cloning conducted by the National Bioethics Advisory Commission. The commission returned in three months time, concluding unanimously that human cloning is "morally unacceptable." The commission also unanimously called for federal legislation to control potentially "unsavory practices" at private research facilities and clinics that bans on federal funding would not reach. Although the commission was unanimous, some members of the commission voiced concerns about Congress's ability to draft legislation that would effectively ban practices that were morally objectionable without encompassing similar research that did not pose similar objections. Steve Holtzman, a commission member, cautioned, "It is very difficult to come up with precise definitions to be sure that you are banning what you intend to ban."[2] Many scientists and ethicists argued that the legislative process was too clumsy a way to respond to the scientific and ethical challenges posed by cloning. Many observers identified federal agencies, such as the NIH or the NSF as better situated to draft the kind of pinpoint regulations that would both prohibit human cloning and permit rigorous pursuit of related medical research.

Ultimately, cloning has become an issue much like flag burning, one which is forever destined to be the subject of political harangues whenever some event brings it into the American public's consciousness. Any actual NIH regulations on cloning—and none have yet been promulgated—are likely to be too complex and technical for the average politician's publicist. In all likelihood, as in other contexts, Congress and congressional staffers will be looking over the shoulder of NIH to ensure that they are on record against the technology. Almost certainly, however, human cloning will take place despite the best efforts of legislators and federal regulators to stop it. Dr. H. Tristram Engelhardt, a professor of medicine and philosophy, noted the basic simplicity of the technology involved, and he commented on the inevitable lessons of this simplicity: "That's why God made offshore islands, so that anybody who wants to do it can have it done."[3]

Crash Test Dummies

Administrative agencies are, of course, subject to myriad pressures from without, ranging from specific interest group lobbying efforts to the more diffuse pressure of the general populace. The National Highway Traffic Safety Administration's (NHTSA) long courtship with air-bag technology illustrates these pressures well. Throughout the 1970s and early 1980s, the Big Three American car manufacturers (Chevrolet, Ford, and Chrysler) lobbied strenuously against any rules that would require air bags, despite overwhelming research indicating that they would save many lives. Only over this intense opposition and many years did NHTSA finally require air bags.

Ironically, however, winning the air-bag war did not put an end to the casualties. Like land-mines exploding long after the hostilities have ceased, the air-bag war continues to injure and maim NHTSA's regulatory efforts. Experience indicates that air bags actually cause some deaths. In particular, and most upsetting, reports indicated that air-bags killed 87 children, including some infants, and small adults in even relatively minor accidents when the bags released with explosive force into their smaller charges. As is often the case, these tragic deaths were accompanied by intense media coverage that served to magnify the scope of the problem many times. Although there was a relatively simple solution to the safety-flaw in the vast majority of cases—placing children in the back seat—these reports of child deaths cast a pall over air bags and led to a public outcry and congressional hearings. Much of this debate focused on whether it should be legal to disable passenger-side air bags so that they would not open in an accident. In the fall of 1997, NHTSA announced a new rule that would give consumers the opportunity to disable air bags if they could demonstrate that small adults were affected or that children necessarily had to sit in the front seat. Somewhat lost in the debate was the fact that, despite these problems, statistics demonstrated that air bags provided extraordinary protection in an accident, so disabling the air bags might contribute to many more deaths than they ever caused. In any case, the air-bag controversy has occupied the substantial attention and resources of NHTSA.

Since virtually all federal research funds on vehicle safety have gone to air bag safety studies,[4] NHTSA has largely ignored other safety issues, including in particular the consequences of crashes between sport utility vehicles and ordinary cars. Yet total casualties of air bags are equaled every *week* in cars that are hit by sport utility vehicles. Although certainly many of these people would have been seriously injured or killed in accidents involving cars, there is a growing research literature indicating that sport utility vehicles are not good for your health if you drive a medium-size car. According to federal statistics, more people die in crashes between light trucks and cars (5,447 last year) than in crashes involving two cars (4,193 last year). This is so despite the fact that, compared to cars, there are half as many light trucks traveling the roadways.

It takes no great flight of the imagination to envision the effects of a collision between a 6,000-pound Chevrolet Suburban and a 2,300-pound Honda Civic. Compounding this weight differential, almost all sport utility vehicles have very high bumpers so that they can be driven "off road," in spite of the fact that only 13 percent of their purchasers ever do so, as shown by a 1995 study conducted jointly by the Big Three auto manufacturers themselves. In a crash, these bumpers clear the main frame (and protection) of a car, with the result that the main impact is on the passenger compartment in a side-impact accident, right over the hood and into the passenger compartment in a head-on impact. Whether it is side impact or head on, the result is very often the same: death to the car occupants and relatively minor injuries for the SUV driver.

Clearly, sport utility vehicles are good for many important political constituencies. Sales reflect strong demand by the public for these vanity vehicles. And certainly in an accident it is better to be in one than in the Ford Taurus you hit. The car companies also love them, since they account for the lion's share of their profits. In addition, under federal law, the minimum standard for fuel economy for cars is 27.5 miles a gallon, but it's only 20.7 miles a gallon for light trucks. This minimum standard is far easier to meet; hence these vehicles contribute heartily to the profits of oil companies, as well.

So far, NHTSA has not conducted sufficient research to promulgate rules that would make light trucks less dangerous to car drivers. Such rules might contemplate weight restrictions, alternative bumper designs, or other precautions. Until research is conducted and rules are promulgated, drivers of cars will be at higher risk.

There are several lessons that can be drawn from this example. Principally, perhaps, if you can afford it, buy a sport utility vehicle. As one accident investigator who had seen one too many deaths in his work explained after he purchased a 4,200-pound Ford Explorer: "Sport utility vehicles are great—for the guy that owns them."[5] Of more enduring importance, however, is the lesson that regulatory bodies sometimes ignore dire statistics and respond to statistics that are less dire but receive considerable attention, such as air bags and their deadly toll on children. This example also illustrates an inherent failure in the market system, a failure in need of regulatory intervention. Drivers of sport utility vehicles have little incentive to change, since they do not incur the full costs of their conduct. Other drivers and society pay for the deaths and pollution caused by these vehicles. Although some of these costs are likely to be internalized in time through higher insurance premiums, most must be addressed through regulation.

Ultimately, responsibility for NHTSA's failure to study the dangers associated with sport utility vehicles lies with Congress. The reasons for Congress's failures in this area are as old as government itself. The Big Three auto manufacturers simply have too much to lose and too much political influence to allow extensive studies that might kill this particular goose. Congress provides little money for crash tests and NHTSA's budget has decreased by 19 percent since 1980, after inflation. In fact, in 1996 and 1997, Congress actually barred NHTSA from spending money to research even whether to raise fuel economy standards, a move that almost certainly would have hurt the sport utility vehicle market.

Congress and federal agencies often play a game of tug-of-war over particular issues. Congress has the authority to halt agency action or to overturn agency rule making simply by passing a law. However, passing a law is not always so simple; it requires majorities of both houses of Congress and the president's signature, which is not usually forthcoming if the law overturns an agency rule promulgated under the president's supervision. As a result, Congress is often reduced to sniping from the periphery. This sniping takes a number of forms ranging from back-room deal making to formal meetings between legislative staffs and administrators.

By far the most popular sniping tactic is the committee hearing, in which agency heads, renowned scientists, and concerned citizens testify before the committee with oversight responsibilities for the respective agency. Although sometimes substantive, a hearing tends to serve two strategic purposes. Foremost, it allows legislators to appear concerned and informed about the matter and provides them with publicity back home. For legislators constantly running for reelection, a controversial committee hearing can provide invaluable exposure, especially if it covers a subject that inspires deep feelings in the legislator's constituents. Additionally, it gives legislators the opportunity to question and sometimes threaten and berate administrators with whom they disagree. This is done either to persuade the administrators of the error of their ways or, more often, just to vent opposition to the agency's course of action.

Who's Afraid of the Big Bad Wolf?

Now this is the Law of the Jungle—as old and as true as the sky;
And the wolf that shall keep it may prosper, but the Wolf that shall break it must die.

—RUDYARD KIPLING (1895)[6]

In 1918, at the behest of Congress and under the auspices of the National Biological Survey, hunters began the systematic extermination of the wolf in Yellowstone National Park.[7] Probably sometime around 1926, the last wolf died or was killed in Yellowstone. The elk, moose, and bison thrived, and ranchers in the surrounding states slept peacefully at night, secure in the knowledge that the Northern Rockies' most efficient predator, after man, was gone. In 1973, however, the gray wolf was listed as an endangered species in the United States under the newly enacted Endangered Species Act (ESA).[8] The bogeyman—bogeywolf?— appeared to be on the verge of returning. The elk, moose, and bison, as well as the rancher and his sheep and cows, could expect many restless nights ahead. Although it took over two decades, finally, in 1995, under the auspices of the National Fish and Wildlife Service, an agency of the Interior Department, a pack of six wolves was set free in Yellowstone.[9] The Yellowstone wolf project was coordinated with a similar reintroduction program in central Idaho. The goal in the greater Yellowstone area was to establish ten packs (around a hundred wolves) by 2002 to reach a sufficiently stable population that the gray wolf could be removed from the endangered species list. At that time, wolf management would be returned to the states.

The story of the gray wolf, especially in the 1980s and 1990s, illustrates the extreme interactions and machinations between administrative agencies and Congress. Like a cacophony resolving itself into a symphony, the wolf's return to Yellowstone illustrates how the system can work.

The Early Years

Under the 1973 Endangered Species Act, the U.S. Fish and Wildlife Service was charged with the responsibility of being the wolf's protector.[10] Under the act, the wolf was listed as "endangered" throughout the continental United States except Minnesota, where it was listed as "threatened."[11] An endangered species is one that is in danger of extinction throughout a substantial part of its habitat. A "threatened" species is one that is likely to become endangered in the foreseeable future. The ESA, however, did not merely mandate the conservation of endangered species; it affirmatively obligated the service to implement wolf recovery. Congress enacted the ESA to promote the "esthetic, ecological, educational, historical, recreational, and scientific value" of the diminishing biosphere.[12] The wolf had been a central component of the Yellowstone biota until the federal government oversaw its extirpation. The federal government, through the Fish And Wildlife Service, now had to restore the ecological balance.

The first recovery plan for the gray wolf in Yellowstone was completed in 1974. Through a combination of administrative lethargy and intense external opposition from the western states and their congressional representatives, no action was taken on this report. In 1983, this now decade-old report was returned to the original drafters so that they could update and revise it. This task was completed in 1987. The Fish and Wildlife Service was "acting" on the mandate established by the ESA, but there was little enthusiasm to complete the task. The gray wolf had become a victim of modern bureaucracy: its fate was being sealed through committee review, extensive evaluation, and thorough study. President Bush's Interior Department continued the policies established in the Reagan years. These two administrations were very friendly to the ranchers and hunters. These particular lambs remembered well Jesus' admonition: "Behold, I send you forth as lambs among wolves."[13] They had no intention of going forth.

Reagan's director of the U.S. Fish and Wildlife Service was Frank Dunkle, a former director of the Montana Department of Fish and Game, which was well known for its dislike of wolves. Dunkle made it clear at the start of his tenure that he did not intend to rush into the matter. Indeed, when Dunkle visited the "Wolves and Humans" exhibit at Yellowstone for the first time he wore a maroon National Rifle Association jacket and cap.[14] Nonetheless, in August of 1987, when the wolf recovery plan reemerged, so as not to dirty his hands with it Dunkle had his assistant sign it. This action appears to have been designed primarily to avoid a lawsuit—since he was obligated to implement the plan under the ESA—rather than any newfound affection for the wolf.

Shortly thereafter, demonstrating a sense of timing and politics that would come to epitomize wolf recovery, a wolf pack in Northern Montana attacked a flock of sheep near the town of Browning. Eight sheep and one lamb were killed. Soon after that, the pack killed three steers. Adding insult to homicide, so to speak, the steers belonged to the same family who had owned the sheep. Some of the offending wolves were killed and some were captured. Whatever other damage they had caused, probably the greatest was to their protectors who sought a home for their brethren in Yellowstone. Once again, the administrative wheels ground to a halt in the face of intense opposition among federal legislators, state governors, and hunters and ranchers throughout the Northern Rockies.

Although wolf recovery was mandated under the ESA, the administrative effort at Fish and Wildlife had now reached an impasse. Congress stepped in to fill the void. In both the House and the Senate, bills were introduced to get the ball rolling and to establish guidelines for wolf recovery. The effort in the House was led by a freshman congressman from Salt Lake City, Wayne Owens.[15] In the Senate, James McClure of Idaho, chairman of the Senate Energy and Natural

Resources Committee, led the legislative effort to have the government reintroduce wolves into Yellowstone.[16] Owens had introduced his House bill because he was a genuine wolf enthusiast and ardent conservationist; McClure introduced his very similar Senate bill because he genuinely disliked wolves and ardently sought a way to protect ranchers and hunters from them. How these conflicting motives could lead to similar proposals reflects the politics and science of wolf recovery as pursued and resisted throughout the years.

Much of the wolf debate was caught up in the mystery of wolf biology. Of course, not all the participants were equally open-minded to what science might provide. Senator Burns (Montana) declared, for instance, "All the science that is used to explain the habits of the wolf will not change my mind to the cold hard facts."[17] Still, all that was known about wolf biology did not amount to very much. The cold hard facts of wolf behavior tended, on close examination, to turn into warm, mushy suppositions. Superstition and speculation, therefore, informed many opinions in the debate. Senator McClure's bill, for instance, was premised on the belief that wolves were migrating, or would migrate soon, into Yellowstone. Once comfortably denned there they would be beneficiaries of the ESA and could not be killed by ranchers even if they were caught in the act of eating a sheep or cow. Hence, McClure, no friend of the wolf, was primarily interested in constructing his law to manage the inevitable wolf return. His proposal allowed for the artificial return of the wolf, but it also would have removed the gray wolf from both the "threatened" and "endangered" species lists on passage. Representative Owens's House bill, in contrast, was premised on the long delay and the expectation that wolves would never reach the greater Yellowstone area without human intervention. He wished to cure the lethargy that had infected the Fish and Wildlife Service. His bill directed the agency to produce an environmental impact statement (EIS), an obligation it already had under ESA, and then move promptly to wolf reintroduction. But Owens, though no friend of the ranchers, also provided for the need to manage this animal whose appetite for predation was never questioned. Owens, after all, was a politician with designs on the Senate and he was from Utah, a state in which the prevailing sentiment is conservatism, not conservation. Owens's proposed solution, which was eventually employed by the Fish and Wildlife Service, was to reintroduce the gray wolves as an "experimental population," an exception to the rigid protection otherwise afforded by the ESA.[18]

Opponents of wolf recovery in Yellowstone were primarily ranchers and hunters. Ranchers were concerned that the wolves would wander out of the park, especially in pursuit of prey leaving the park in winter, and would feast on their livestock. Representative Marlenee (Montana), a vocal opponent of wolves, described in vivid detail and with accompanying pictures how wolves could be expected to rip apart and eat the innards of "Bessie, the milk cow."[19] Senator Burns warned that "the fact [is] that these animals [have] killing on their minds. They will seek out the easiest course of action. . . . This is an animal that enjoys lamb, mutton and beef as well as you or I."[20] He also provided pictures. Hunters, on the other hand, were afraid of what the wolf would do to the ungulate populations, primarily elk and moose, in the area. Hunters were already frustrated with the hunting prohibition inside Yellowstone at a time when the ungulate populations were soaring. The hunters were forced to wait until the herds wandered off federal property. The herds, however, did not always cooperate, as Interior Secretary Bruce Babbitt explained: "The elk . . . and the bison are pretty smart critters. They can't read, but they know where the guys with the guns are, and they tend to bunch up inside the park boundaries during migration seasons."[21] Now the wolves would exacerbate the problem. Representative Hansen (Utah), a strong wolf opponent, described the wolf as "a shark on dry ground."[22] Hunters and the states that benefited from their presence expected wolves to feast on the Yellowstone ungulates and argued that every elk eaten by a wolf would be one less elk they could shoot. These predators knew a competitor when they saw one.

Wolf Biology

The public debate before both the Fish and Wildlife Service and various congressional committees was profoundly muddled. Although all sides rested their conclusions on biological premises, few of the debaters could agree on what these were or what might result from wolf reintroduction. Everyone agreed that wolves were predators who ate ungulates and that they would find a bounty of prey in and around Yellowstone. Everyone also agreed that, although essentially territorial, wolves roam widely and should be expected to wander off federal property and onto state or private property. Wolves have been known to travel more than 500 miles in a week. Proponents acceded that the wolves were not very likely to stay inside Yellowstone. Senator Burns pointed out that "wolves don't know where the boundaries are, and they don't read."[23] Representative Young (Alaska) warned similarly that the wolf "is not an animal that lives by rules, nor can he read."[24] He went on to predict direly that "he will go on beyond the borders of the Yellowstone Park and he will get into the cattle rancher's area and into the sheep rancher's area, and even downtown, possibly, in some of the communities."[25] What the wolf would *do* to the ungulate populations, or to livestock, or even in downtown, however, was the subject of much disagreement. The story of the wolf in Yellowstone is a story of policy making in the face of significant biological uncertainty.

Both sides did agree about some of the particulars of wolf biology, though even then they tended to disagree about the meaning of this biology for policy formation. Foremost, everyone apparently agreed that wolves don't read. Both sides also agreed that there were approximately 60,000 gray wolves in Canada, 8,000 in Alaska and 2,000 in Minnesota, Wisconsin, and Michigan. But the two sides feverishly debated the significance of these numbers. Since the ESA lists species by region, the gray wolf was listed as "endangered" in the Yellowstone area despite these numbers. This region by region evaluation of species was added to the ESA in a 1978 amendment. This approach to conservation treated "distinct population segments" separately for purposes of federal law, thus having the anomalous effect of labeling a species "endangered" despite the large numbers inhabiting other regions. For opponents, this was not anomalous, it was perverse. Opponents believed that these large numbers indicated that wolves were hardly in need of federal protection and, moreover, that there was something desperately wrong with a federal law that protected them. A General Accounting Office report also registered concern about the 1978 amendment, speculating that squirrels in a city park might be designated a distinct population segment.[26] Senator Malcolm Wallop (Wyoming), commenting on the odd categorizations that protected the wolf under the ESA, concluded, "To quote a famous American, the law is an ass."[27] As Charles Dickens, a famous Englishman, might have observed about Wallop, "He has gone to the demnition bow-wows."[28]

Proponents rested on both the technical letter of the law, which accorded wolves protection, and more substantially on the ground that putting the wolf back in Yellowstone would restore the ecological balance upset when the federal government supervised the wolf's extermination in the 1920s. In particular, the elk and moose herds in Yellowstone had gone unchecked and were now grazing out of control and causing substantial soil erosion problems, since their natural predator, the wolf, had been removed.

The antagonists in the wolf debate ultimately disagreed about most of the details of wolf biology and about what could be expected empirically on its return to the neighborhood. Indeed, they did not even agree on the historical matter of whether wolves were natural to Yellowstone and thus whether the program was a "reintroduction" or an "introduction" of the gray wolf into the area. Some opponents argued that wolves were not found there prior to 1910 and thus the entire foundation for the program was flawed.[29] As the debate heated up, the disputants became more entrenched in their positions on wolf biology. Here are some of the questions, together with the factual disagreements, that plagued virtually every discussion of wolf reintroduction. You will notice that consistency was not highly regarded by either side in the debate.

1. **Do wolves engage in surplus killing?**

   OPPONENTS: Wolves engage in surplus killing, killing more than they need to eat, continually moving on to the next hapless victim. One rancher commented hysterically: "It won't be long before there are hundreds of wolves over thousands of square miles, killing for fun."[30]

   PROPONENTS: Wolves do not engage in excessive surplus killing and certainly not at a level that would endanger the herds of elk in Yellowstone, which exceed 30,000 head.

2. **What impact will wolves have on the breeding strength of the Yellowstone ungulates?**

   OPPONENTS: Wolves will have a devastating impact on elk and moose populations because they naturally prey on the strong and fit animals.

   PROPONENTS: Wolves will have a positive effect on the elk and moose populations by weeding out the weak and diseased animals and thereby evolutionarily strengthening the breeding stock of these populations.

3. **What impact will the wolves have on the numbers of ungulates in Yellowstone?**

   OPPONENTS: Wolves will not cull the inflated elk, moose, and bison herds sufficiently; the government should permit hunting in Yellowstone as a more effective conservation measure.

   PROPONENTS: Wolves will cull the now too large elk, moose, and bison herds and thus be an efficient conservation measure.

4. **Will wolves prey on livestock and domestic animals?**

   OPPONENTS: Wolves eat (indeed prefer) sheep and cows and will kill domestic pets. Their predation has been a constant problem for the ranchers of Minnesota.

   PROPONENTS: Wolves are not inclined to attack domestic animals, especially when there is plentiful prey to be found, as is the case in and around Yellowstone. In Minnesota, actual wolf predation on livestock is only about one-tenth of 1 percent.

5. **Will wolves significantly reduce hunting success in the affected states?**

   OPPONENTS: The wolf will "decimate" the elk and deer populations and cost the states millions in lost revenues due to diminished hunting stocks.

   PROPONENTS: Wolves are unlikely to have any appreciable effect on hunting success; in fact, experience in Minnesota suggests that even a combination of a harsh winter and wolf presence did not reduce hunting success greater than 5 percent, and then only for a brief period of time.

6. **What impact will the wolf have on Yellowstone's other signature endangered species, the grizzly bear?**

   OPPONENTS: The wolf is bad for the grizzly bear because the wolf will eat too many of the bear's natural prey and might also eat grizzly cubs.

   PROPONENTS: The wolf is good for the grizzly bear because surplus kills will be left for bears, especially benefiting them when they emerge from hibernation.

7. **What are the costs of the program?**

   OPPONENTS: The actual out-of-pocket costs are estimated at over $1 million per wolf; in addition, the presence of the wolf will costs tens of millions (and possibly hundreds of millions) due to decreased hunting, reduced tourism (because areas will be closed to protect wolf habitat), and losses of livestock and domestic pets.

   PROPONENTS: The actual costs of the program will be about $12 million through the year 2002; hunting will not be affected, tourism will increase and thus increase revenue to surrounding states and National parks, and losses due to depredation will be covered by a private fund.

8. **Would the wolves return naturally to the Yellowstone area from points north where they are abundant?**

   OPPONENTS: Wolves are already recolonizing the greater Yellowstone area, so what is needed is control of "these critters," not more of them.

   PROPONENTS: Wolves are not recolonizing and will not recolonize the greater Yellowstone area.

9. **Will artificially introduced wolves taken from Canada simply migrate back home?**

OPPONENTS: Any wolves artificially introduced into Yellowstone from Canada will simply turn around and head home at their first opportunity; the enormous sums spent on this project will be wasted.

PROPONENTS: The wolves will not migrate home to Canada, especially if they are slowly habituated to the area around Yellowstone.

10. **Is the gray wolf really a wolf, or is it a coyote-wolf half-breed?**

OPPONENTS: DNA tests indicate that the modern gray wolf interbreeds with coyotes and is thus not the "species" that is protected under ESA, nor should it or can it be so protected.

PROPONENTS: Although DNA tests indicate that interbreeding has occurred, it is not usual and appears to have occurred in the distant past and only in a very isolated manner; in general, wolves displace or kill coyotes.

11. **Is there public support for wolf restoration?**

OPPONENTS: Public opinion polls demonstrate little support for the wolf.

PROPONENTS: Public opinion polls demonstrate strong support for the wolf.

12. **Reflecting on the start of the program (in May 1995), has it so far been a success?**

OPPONENTS: No, it's a major failure.

PROPONENTS: Yes, it's an outstanding success.

The Later Years

The wolf's prospects in Yellowstone brightened dramatically in the executive department with the election of Bill Clinton in November 1992 and, in particular, his appointment of Bruce Babbitt as secretary of the Interior. Babbitt was deeply romantic about returning the wolf to Yellowstone, believing it to be necessary for restoring the balance and character of a park that he referred to as an "American Serengeti."[31] This contrasted with the late 1980s and early 1990s, when Congress had spurred the executive branch to action on wolf recovery. In 1988, they directed Fish and Wildlife to conduct extensive studies on reintroduction. In 1990, Congress created the Wolf Management Committee to devise a concrete plan that would be acceptable to the myriad interests involved in this issue.[32] In 1991, Congress directed Fish and Wildlife to prepare an environmental impact statement.

The wolf's prospects dimmed considerably, however, at the midterm elections in 1994 for the 104th Congress. This "revolution" ushered in Republican majorities in both the Senate and the House of Representatives who were dedicated to balancing the budget and whose loyalties favored the ranchers and hunters. The biggest House wolf proponent, Wayne Owens, had given up his seat in 1992 in a failed attempt for a Senate seat. Representative Barbara Cubin (Wyoming) now warned that the choice was between wolves on the one hand and children, the sick, and the elderly on the other. She concluded, "Wolves in Wyoming simply cannot compare to these priorities no matter how good the howl of the wild makes some people feel."[33]

The congressional hearings on wolf reintroduction turned decidedly cooler to wolf proponents—and they had not been terribly temperate from the start. But now the wolf had good friends in the highest reaches of the federal bureaucracy. With Bruce Babbitt, the wolf now had the Interior Department squarely on its side.

Just six days after the "revolutionary" 104th Congress started its session in January 1995, the Fish and Wildlife Service released a total of twenty-nine wolves in two states. After nearly two decades of inaction, the federal bureaucracy thus demonstrated an extraordinary efficiency in completing the task. In response to pointed congressional questioning, however, Secretary Babbitt specifically denied "hurrying up" the release.[34] And perhaps the timing was merely coincidental. But there is no question that Congress would find it more difficult to stop the process now that it had begun. This fact was clearly evident in the congressional hearings held immediately following the release, in which Secretary Babbitt was accused of a series of transgressions ranging from violating the Tenth Amendment by ignoring the affected states' objections to wolf reintroduction to violating the Establishment Clause of the First Amendment by sponsoring an ecoterrorist religion. Republican Representative Helen Chenoweth (Idaho), who during her election campaign had questioned the endangered status of salmon in light of the fact that she could buy it canned in her local supermarket,[35] fumed, "This would be a giant step closer to the utopia religious environmentalists are striving to create— a utopia where human beings have only as much value as the razorback sucker fish, and possibly less."[36]

The Compromise

The details and implementation of the wolf reintroduction plan turned out to be a model of reasonableness. It was a solution that gave neither side all it wished for, nor did the plan impose on either side all it feared.

It was thus a model of political compromise. The plan had two essential features: close management of the reintroduced wolves and compensation to ranchers for any wolf depredations that occurred.

A key element of the plan was section 10(j) of the ESA, which permitted the Fish and Wildlife Service to designate the Yellowstone wolves a "nonessential experimental population." This was pivotal to the success of the program since it allowed a much more flexible response than was permitted while the wolves remained "listed" as endangered. Under section 10(j) the Yellowstone wolves, as a separate breeding group in an area not shared by other listed wolves, could be fully managed without adversely affecting the species generally. Indeed, under the law, section 10(j) could be employed only on a finding that this approach would benefit "the conservation of the species."[37] Section 10(j) provided public officials and private landowners much greater flexibility to respond to the possibly more egregious predatory actions of the wolves. For instance, under this protocol, wolves could be harassed away from private property and could be killed if caught "in the act of killing livestock."

Not surprisingly, this compromise raised the ire of both extremes represented in the dispute. Many conservation groups, finding themselves awkwardly aligned with ranchers in various lawsuits, believed the experimental designation was too permissive. Ronald Snodgras, for instance, the Audubon Society's wildlife policy director, complained that the experimental designation made no sense: "We'll never get them off the [Endangered Species] list if we allow government agents and stockmen to kill wolves."[38] Groups like the Audubon Society and the Sierra Club opposed the prospect of wolves being killed on principle and refused to consider compromising to the political realities of the situation. Ranchers, in contrast, complained that the experimental designation was not permissive enough. Killing wolves when they are caught in the act is rather more difficult to do in practice than it might sound in theory. Bob Budd, of the Wyoming Stockgrowers Association, observed that "wolves generally don't come down and punch a clock and say 'OK we're gonna kill your cows now.'"[39]

The limitations associated with section 10(j), however, did not outweigh the clear benefits gained by designating the Yellowstone wolves as a nonessential experimental population. The ranchers could not declare open season on the wolf and, at the same time, some wolves could be "taken," the bureaucratic euphemism for killed.

It turns out that a more substantial objection to section 10(j) was the statutory requirement that natural populations be kept separate from experimental populations. This would be no problem so long as there were no wolves in Yellowstone and none would be migrating there for the duration of the program. Complicating matters considerably, however, was an incident in the fall of 1992 in which a hunter, Jerry Kysar, shot and killed an animal he initially took to be a coyote in the Teton wilderness just outside Yellowstone. On closer inspection, however, it turned out that Kysar had killed a wolf. This stirred up a lot of dust and it threatened to derail the service's plans entirely. Indeed, about a week after Kysar killed his wolf, Yellowstone rangers reported a sighting of an entire pack in the same area. However, after extensive tracking and other monitoring turned up no evidence of other wolves, the service concluded that Kysar had shot a lone wolf that had wandered 400 miles from Glacier National Park in Montana. The 10(j) designation was secure. Under the law, only a breeding population would interfere with the section 10(j) plan. As Ed Bangs, the service's wolf project leader noted dryly, "I say, unless my biology's off, it still takes two."[40]

The other pivotal political aspect of wolf reintroduction was creation of a fund to compensate ranchers who lost livestock to wolves. In 1987, the Defenders of Wildlife initiated their Wolf Compensation Fund. This fund would be entirely private, thus not costing American taxpayers one penny. The fund, with over $100,000 committed to it, would pay the market value of any livestock killed by a wolf. In addition, Defenders of Wildlife promised $5,000 to any rancher whose property was used by wolves for denning. Ranchers complained, however, about the forensic and bureaucratic difficulties associated with actually demonstrating that a particular cow or sheep had been killed by wolves. But there was little basis for complaint. Prior to the implementation of reintroduction in Yellowstone, the Fund had paid out over $12,000 to ranchers in Montana claiming losses to wolves. Moreover, the administrators of the fund consistently erred on the side of giving the benefit of the doubt to the claimant if there was disagreement about the identity of the predator. As proponents had certainly hoped, the existence of this fund put a substantial damper on the key argument of ranchers that they would lose property as a consequence of wolf reintroduction. Ironically, in the end, ranchers were reduced to the complaint that the fund was private and that Congress should guarantee compensation with public funds, an uncomfortable position to take for people accustomed to arguing that taxes and government should be reduced.

When Politics and Science Meet

The decision to put wolves back in Yellowstone was ultimately a political decision that involved substantial compromises by all sides. Although wolf reintroduction was informed by science, our understanding of the biology of wolves was never enough to drive either the debate or the decision. Representative Hansen, who had opposed reintroduction, lamented after the first wolves had already set foot in Yellowstone, "I always worry though that sometimes we don't come up with the science that is necessary to back up what we are saying. Just because we feel strongly about it and we have the burning in our bosom is hardly the criterion of whether we should do it."[41] But even the best science available could only offer a broad template for organizing the principles, prejudices, and passions surrounding the wolf. If science was never irrelevant and was even instrumental to all that occurred over the years, there never was a substantial body of research available on the most pressing issues of wolf reintroduction. Yet even if more work had been done on wolves, it probably would not have contributed to the debate substantially. Wolf biology was too complex and the ecosystem in which they were being put was unlike any previously studied. The scientists could never entirely reassure the ranchers and hunters, nor could they entirely endorse the conservationists' romantic vision of wolves. All they could say was that the wolves were unlikely to be as voracious as feared by some people, although some wolves might be voracious at times. The science paled in comparison to the passions that drove the dispute.

In the heat of the debate, one rancher exclaimed that "to hear the howl of the wolf is a pretty lame reason to bring them back." Secretary Babbitt, in contrast, observed that the wolf restored the balance to Yellowstone and its howl was evidence of this harmony. Opinion on wolf introduction pretty much depended on one's opinion of the wolf's howl. It was thus largely a value choice, and science played only a supporting role in the drama. Wilderness philosopher Aldo Leopold perhaps summed up the lesson best: "Only the mountain has lived long enough to listen objectively to the howl of a wolf."[42]

Blending Science with Policy and Politics

Get your facts first, and then you can distort them as much as you please.

—MARK TWAIN

The story of the wolf's return to Yellowstone provides an enduring lesson of the use of science in the administrative process. Policy and political considerations play an enormous role in how scientific research is digested by those charged with making the decisions. Special interest groups and the agencies themselves have specific agendas that might or might not be affected by the results of science. In theory, science should be free of political bias and should be used as a tool of enlightened decision making. In practice, science rarely plays such an exalted role.

Policy and politics, of course, are not the same thing. Policy refers to the values or norms that underlie and guide political decision making. Politics is the process by which these values are integrated into the decision-making process. Hence, the goal might be to preserve the spotted owl, and politics is the battlefield on which the war is fought. Yet while politics is often simply self-interest, this is not to say that norms are not involved. The norm might be reelection rather than species preservation. Moreover, since a candidate rarely expresses a policy preference of self-promotion candidly, he or she is likely to embrace a substantive policy view. In such cases, therefore, the candidate is likely to advocate either the spotted owls' or the loggers' position even if this preference is based on self-preservation rather than preservation of owls or jobs.

The main issue to be considered here, then, is how norms and principles—what I will refer to as policy preferences—interact with scientific understanding of the facts. Whether they are for good or ill, policy preferences shape and sometimes overwhelm decisions ostensibly based on science. Since policy and science lie in uneasy tension, often fully overlapping in practice, so their relationship is not easily disentangled.

Policy preferences also play a significant part in the actual conduct of the scientific enterprise. From the hypothesis chosen for study and the design of the study to the statistical analyses run and the conclusion inferred from the results, the scientific method is replete with value choices. Scientists, of course, are intimately familiar with these choices and attempt to avoid or temper them at every turn. A good example of this is the double-blind research method, in which both subjects and experimenters do not know, during the data-gathering process, which are the experimental groups and which the control groups. This mechanism ensures that experimenters do not simply observe what they expect and subjects do not perform as expected. Another control mechanism is replication, whereby researchers, possibly with different values or agendas, attempt to repeat another group's findings. One way to understand the definition of the scientific method is as the systematic attempt to remove all bias from the study of testable factual questions. Science never succeeds fully in this goal, but its value can be measured largely by how near it gets.

No matter how pure the science, it's up to us to figure out what to do with it. This is true no matter how unambiguous the results. For instance, few seriously doubt that cigarette smoking increases the likelihood of dying a premature death. Although most people might be led not to smoke on the basis of this science, it is not the science that compels that result. The factual connection between smoking and premature death merely describes what is so; policy preferences must tell us what we ought to do. Science, however, does not always provide an unambiguous picture of the world; and the fuzzier the view science gives us, the greater latitude there is for policy to complete the picture.

In general, it is probably accurate to say that the more certain the science, the less room there is for policy preferences. Just to offer a straightforward example, suppose the issue concerns NASA's preparations for identifying and destroying large meteors that might strike the earth. As the dinosaurs apparently discovered, a large meteor strike can have devastating effects on the earth and its inhabitants. But NASA, unlike Hollywood, spends little of its budget on this contingency, since, despite its cataclysmic potential, the likelihood of such an event is presumed to be exceedingly remote, in spite of the fact that NASA has never studied the matter very closely. Our policy is to spend money on other priorities, and this leads us, together with hopeful thinking, to discount the likelihood. If, however, scientists identified a meteor on a collision course with the earth with a high degree of certainty, there would probably be little policy debate on what to do. While undoubtedly some people would welcome the meteor as our biblically preordained destiny, NASA and most of the federal government could be expected to invest considerable sums in trying to avoid this fate.

But the realities of the science and politics relationship are rather more complex. First of all, on the science side, the uncertainty associated with research concerns not only questions regarding the likelihood that the conclusions are true but their significance if they turn out to be true. Returning to our meteor, there are really two scientific issues. The first is the likelihood that the earth will be hit by a meteor, and the second is how much damage it would cause if it did hit. The real-world meteor of today is the greenhouse effect and the global warming expected from the shroud of carbon dioxide that we have thrown over our planet. Every discussion of global warming comes down to a debate over the merits of the science. Most scientists believe that temperatures will rise. Consensus breaks down, however, on where they will stop. In particular, scientists disagree about whether the warming will be three degrees or eight degrees centigrade over the next century, with the latter estimate positing catastrophic flooding of low-lying areas. Whether you believe we should spend billions to slow this effect depends largely on your trust in the numbers and, indeed, whether you believe that it is not too late to reverse this warming trend.

On the policy side, the decision process also will be affected by the degree of zealotry felt by the contending parties. The more strongly the political outcome is desired, the more likely the science, however certain it might be, will be manipulated, ignored, or rebutted. If, for example, some industry groups were significantly disadvantaged by NASA's efforts at meteor intervention, they might be expected to challenge the science or offer industry-sponsored studies to rebut the government's research. Thus, industries asked to foot the bill might emphasize just how little is known about killer meteors and the devastating impact on the economy of spending billions to avoid meteors. Maybe dinosaur extinction was not really a consequence of a meteor impact. (Gary Larson's *Far Side* cartoon, for instance, suggested an alternative possibility. Pictured were dinosaurs standing around smoking cigarettes. The caption read, "The real reason dinosaurs went extinct.") This is exactly the response industry groups have had to the prospect of global warming. To be sure, the costs associated with avoiding the predicted warming are astronomical, so it is reasonable to demand a sound scientific basis before spending vast sums. At the same time, the consequences are sufficiently dire that we might not want to wait too long.

An interesting aspect of the zealotry component of politics concerns the manner in which the debate takes place. Those with very strong policy views tend to fight the science in two very different ways. The first way, which is especially effective if the science is not overwhelming, is to attack the methods and possibly the scientists themselves behind the unwelcome results. This strategy sometimes includes offering contrary research or inconsistent anecdotal reports. Generally, this strategy is not entirely unwelcome, since scientists and their methods should be held to the highest standards. But sometimes opponents use this strategy to obfuscate and beat down the reputable science for the single purpose of killing it. The classic example is the surprisingly effective tobacco industry effort to undermine the prodigious literature indicating the deleterious health effects of smoking. Despite overwhelming evidence to the contrary, using science-speak itself, the tobacco industry was able to raise doubts about the causal connection between cigarettes and lung cancer. Of course, in the end, albeit billions of dollars in sales later, the science was just too strong to rebut.

The second strategy employed, especially to confront very strong but unwelcome scientific news, is to challenge the relevance of the science entirely. The tobacco industry used this strategy when it turned to the principle of "individual liberty" to rebut the scientific challenge. This is an appealing argument to the average American citizen/consumer: smoking may (or may not) be bad for you, but government should leave you alone to decide for yourself. The government's very predictable response to tobacco's newfound appreciation for the Bill of Rights was to change the ground on which the debate took place. The government focused attention on the children. The liberty argument was largely irrelevant when applied to them, and the science basis was again moved to center stage.

An even more vehement variation of the rejection of the relevance of science occurs when groups challenge the value of the science paradigm entirely. Many oppose arguments premised on scientific research by challenging the "privileged" status of science. They claim that the "western scientific tradition" is a hegemonic discipline and that there are many ways of knowing the world that are just as legitimate, if not more so, as the scientific method. There are at least two versions of this antiscience perspective. The more radical holds that there is no reality at all. The world that is seemingly out there might be nothing more than a product of our imaginations. (Or, more accurately, *my* imagination, since I cannot be sure that you are not simply a creation of my mind.) The less radical version holds that there is something out there, but what that something is can be known only subjectively or individually. "Truth," therefore, is socially constructed and can change with passing fashions and according to which group happens to be in power. It is hard to know how seriously to take the radical version of the scientific critique. I suspect that even the most hard-core adherents to it look both ways before crossing the street. The less radical version, generally associated with the continental school of philosophy and philosophical deconstructionism—and what I will refer to collectively as "postmodernism"—has had a fair amount of success in the legal academy. It is today associated with critical legal studies, critical race theory, and critical feminist scholarship. To the extent that postmodernism competes with scientific realism for adherents, which it does, it deserves some attention.

In the spring of 1996, heavy rains flooded the Columbia River in Kennewick, Washington, and revealed a long-buried skeleton. Soon after it was discovered by a couple of college students, the skeleton came to the attention of Jim Chatters, a forensic anthropologist. Using radiocarbon testing, Chatters estimated that the skeleton was approximately 9,300 years old. He dubbed the skeleton "Kennewick Man." Of further relevance to my story, Chatters determined that the skeleton had Caucasoid features that linked it to regions of Europe and southern Asia. Native American tribes, in contrast, have northern Asian features. Therefore, Chatters suggested, this skeleton might indicate that some of the original settlers of North America were from southern Asia or Europe rather than, as long believed by scientists, northern Asia. Intrigued by these observations, he wished to study the skeleton further.

The Army Corps of Engineers, however, at the behest of several Indian tribes, confiscated the skeleton. The Indian tribes sought possession of Kennewick Man so that they could give it a proper burial. The corps moved the skeleton to a government storage facility until the dispute could be resolved. The Indians claimed sovereignty over the skeleton under a 1990 act, the Native American Grave Protection and Repatriation Act (NAGPRA). NAGPRA provides for the repatriation of cultural property to its rightful owners. The Indian tribes claimed that the skeleton, if it is 9,300 years old, is necessarily part of their cultural ancestry and within the provisions of the statute. Chatters and other scientists contended that since the skeleton is Caucasoid, it is not related to the groups claiming kinship with it.

There are two competing versions of the "truth" that, in this case, a court will have to resolve. The scientists argue that the skeleton is the "ultimate elder," and since it predates the Indian tribes' arrival in the area, it cannot be "repatriated" within the meaning of the statute. This claim rests on the science of radiocarbon dating. The Indian tribes, in contrast, date their existence back more than 10,000 years through their narrative history and thus claim to be the skeleton's heirs. As an Indian spokesperson explained,

> From our oral histories, we know that our people have been part of this land since the beginning of time. We do not believe our people migrated here from another continent, as the scientists do. We do not agree with the notion that this individual is Caucasian.[43]

Consequently, if the skeleton is 9,300 years old it is, by definition, an Indian ancestor, since they have been here from "the beginning of time." The fight over Kennewick Man, therefore, appears to offer a contemporary version of the struggle between evolutionary theory and creationism.

In the context of the debate between postmodernism and science, the fight over Kennewick Man offers a variety of additional lessons beyond the basic choice between faith and reason. First of all, postmodernists wisely avoid religious debates, correctly thinking of religion as potentially as hegemonic as science. Nonetheless, postmodernists easily rally to the narrative perspective of the Indian tribes, since they too extol the virtues of narrative and seek to champion groups historically disadvantaged by the dominant hegemony. In addition, in the context of Kennewick Man, the choice between faith and reason is somewhat muddled, a state of affairs that postmodernists seem to revel in. Specifically, it is not necessary to reject the viability of science in order to decide this matter in favor of the Indian tribes. The science, quite simply, might be irrelevant given the applicable value judgments embraced by the statutory scheme in these cases.

NAGPRA was enacted largely to address and remedy the tragedy associated with the pillaging of Indian grave sites in the nineteenth and early twentieth centuries. This remedial purpose might require judicial deference to Indian sensibilities. If this is so, the Indian narrative prevails not because it is more "accurate" than the scientists' conclusions. It prevails because the scientists' truth is not relevant under the pertinent law. I don't need to take a position on the merits of the dispute to fully appreciate the reasonableness of interpreting NAGPRA to require deference to the Native American viewpoint.

What would not be reasonable, however, would be a modern court ruling for the Indian tribes because it believed their narrative and did not believe the scientists' data. Postmodernists adopt this tenuous position, arguing earnestly that a scientist's "narrative" is no better at discovering the "real story" than any other. Hence, for them, there is no difference between a court's accepting the accuracy of the Indians' version of truth and the scientists' conception of truth. All narratives are equal and none should be privileged. In my view, the Indians' narrative might be privileged here, but it must be based on an explicit value choice, as might be enumerated in the applicable statute. But when it comes to scientific verity, there is only the scientists' version.

Since courts are somewhat reluctant to accept narratives in preference to radiocarbon dating, more practical-minded postmodernists— which itself might be a contradiction in terms—sometimes dress up their narratives in the guise of science. Although this has not happened in the case of Kennewick Man, there are plenty of examples. The most obvious is in Kennewick Man's analogue, evolution. If the priests had not invented it, creation science would have been a perfect postmodernist tool. Other examples we have already seen are battered woman syndrome and rape trauma syndrome. Although not invented by postmodernists, these pseudoscientific theories are ardently advanced by them in order to accomplish their normative goals. For postmodernists, all science is storytelling. The only question is who can tell the better story.

In the usual agency decision, the science is not known with anything approaching the degree of confidence that we have for radiocarbon dating or the theory of evolution. Hence, in most situations, administrators must make policy decisions under conditions of more or less scientific uncertainty. Under such circumstances, there is substantial leeway to massage the science in order to achieve desired normative outcomes. One need not be a postmodernist to have doubts about the value of the science in these cases. Too often the science is made to appear more certain than it is in order to hide the policy preferences driving the decision. A good example of this tendency comes from our generally successful efforts to clean the air.

Clearing the Air

Under the Clean Air Act, the Environmental Protection Agency (EPA) is charged with setting national standards for certain air pollutants to protect public health and the environment.[44] Indeed, EPA's statutory mandate, or mission, is quite broad: "to protect human health and to safeguard the natural environment—air, water and land—upon which life depends." Among numerous other responsibilities, EPA regulates nearly 200 toxic air pollutants emitted from industrial facilities, including benzene, chromium, cadmium, and vinyl chloride. EPA also is concerned with air visibility and regulates the pollutants that produce the haze that impairs visibility. In order to improve EPA's monitoring of air quality, since 1977 Congress has required EPA to conduct formal reviews every five years of national ambient air quality standards (NAAQS). This review was designed to promote public health by having air quality standards that are based on the latest scientific findings. Unfortunately, science does not always fit snugly in the policy hole set for it. But as every child who has tried to fit a square peg into a round hole eventually learns, you can do it if you force it.

In 1982, EPA conducted a review of the particulate matter (PM) criteria and revised the NAAQS for PM in 1987. However, EPA failed to review the standard again in 1992, as required by law. In a federal district court in Arizona, the American Lung Association sued EPA, seeking to force the agency to comply with the law.[45] The American Lung Association claimed that recent studies indicated that the current particulate matter standard of PM_{10} was too lax and endangered public health.

Particulate matter refers to a mixture of solid particles and liquid droplets found in the air. These airborne particles come in a variety of sizes measured in micrometers. A million particles, each ten micrometers in diameter, or PM_{10}, could fit on the head of a pin. Different-size particles tend to come from different sources. Coarse particles, larger than 2.5 micrometers, come from such sources as windblown dust and grinding operations involved in roadwork. Fine particles, less than 2.5 micrometers, come from fuel combustion, power plants, and diesel-powered vehicles. With the 1987 revision, EPA adopted a national standard of PM_{10}. According to EPA, the best research available at that time indicated significant health effects associated with concentrations of particles up to ten micrometers in diameter. Some of the health consequences associated with PM included aggravated asthma and chronic bronchitis, with the young and the elderly most at risk. Under the regulations, PM would be monitored both over certain twenty-four-hour periods and annually. EPA did not return to the PM_{10} standard until ordered to do so by the Arizona federal court in 1994.

Although spurred by a court order, the administrators at EPA and especially its head, Carol Browner, soon became true believers in the process of review and in the need to update PM standards. In an interesting and possibly bizarre twist to the NAAQS story, EPA decided to tie another pollutant, ground ozone, to its PM process. This tie-in arrangement was like going to a restaurant and ordering a steak and having to pay for the broccoli that came with it whether you wanted the broccoli or not. The difference is that EPA has the regulatory authority to make you eat the broccoli.

Ozone, without question, is a major pollutant that has long appeared on EPA's radar screens. Ozone is the principal ingredient in smog and historically has been considered the leading air quality problem in the United States. In the stratosphere, ozone provides a life-sustaining screen against the sun's damaging rays. But at ground level, though it continues to have some beneficial effects, ozone is associated with myriad deleterious health effects, ranging from aggravating asthma to impairment of the body's immune system contributing to respiratory illnesses.

EPA had last revised its ozone standard in 1979. The regulation set the acceptable limit at 0.12 parts per million (ppm), meaning that an area could not exceed this level as measured by a continuous ambient air monitor more than once per year when averaged over three consecutive years. EPA completed an extensive reassessment of its ozone standard in May 1989 and reaffirmed the standard in March 1993. Yet just three years later, EPA announced proposed revisions for ozone that it then coupled with its PM proposals. The decision to couple ozone with PM struck many observers as a disingenuous political move, not at all based on the scientific merits. As Representative Ron Klink (Pennsylvania), a strident critic of the new EPA guidelines, observed: "When you look at the ozone standard alone, the EPA's own cost-benefit analysis shows that the costs outweigh the benefits of the new standard. But when combined with the PM standard, that is no longer true. It may be that the ozone and $PM_{2.5}$ standards were linked to hide the cost of the ozone standard."[46]

In reviewing the national standards for certain air pollutants, Congress requires EPA to consult with an independent body of scientists known as the Clean Air Scientific Advisory Committee (CASAC). Although EPA has access to a bevy of scientific advisors inside the agency, CASAC was created in order to provide an independent outside review of newly proposed standards. EPA, however, has no legal obligation to follow any CASAC recommendation. In practice, CASAC exists as much to provide Congress and EPA's critics with grounds on which to challenge proposed regulations as it does to give EPA guidance. In that capacity, CASAC serves as an integral part of the checks and balances existing between the branches of the federal government.

In reaching their conclusions on PM and ozone, Carol Browner explained that EPA and CASAC had waded through "thousands of peer-reviewed scientific studies."[47] Despite this impressive-sounding bounty of data, the CASAC scientists were deeply divided about what prescriptions the research might commend. The scientists did agree on several broad policy recommendations. For instance, they unanimously agreed that the ozone-measuring standards should be changed from a one-hour averaging period to an eight-hour period. The longer evaluation period would provide a more accurate exposure picture. In addition, a majority supported adoption of a fine particle standard, for particles smaller than 2.5. An increasing confluence of research indicated that there were substantial health risks associated with particles smaller than the current standard of PM_{10}. These "fine particles" might pose special risks because they were inhaled deeply into the lungs. The scientists, however, divided sharply over the appropriate ranges or levels at which these fine particles should be monitored. This left EPA largely free to choose among the wide range of numbers alternatively embraced by different members of CASAC.

Dr. George Wolff, chairman of CASAC's panel on ozone and PM, explained in testimony before a House subcommittee that CASAC did not endorse the levels and standards selected by EPA. Although most CASAC members agreed that a new size of $PM_{2.5}$ should be established, the members sharply disagreed on how this standard should be implemented. Wolff explained that there were essentially four basic views on this issue. Only four panel members advocated measurement standards for $PM_{2.5}$ at or near the lower end of EPA's recommended ranges. Seven members recommended levels just above EPA's ranges. Eight members refused to identify any specific ranges, though they endorsed the $PM_{2.5}$ level. The remaining two members did not believe a $PM_{2.5}$ standard was warranted at all.[48] Although EPA's proposed guidelines were thus "consistent" with CASAC's recommendations, they were uniformly more stringent than those most CASAC members embraced. CASAC concluded that the science was too weak to lend support for any of the *specific* levels chosen by EPA.

At its best, establishing a standard that incorporates good scientific research with considered policy judgments is a high-wire act. As CASAC's report makes plain, EPA walked this particular wire without a safety net. The science was not there to save EPA if it fell. Just considering PM alone—arguably the better supported of the two new guidelines— CASAC identified a litany of weaknesses and uncertainties associated with the studies it reviewed. Just a few of them included unexamined confounding variables, untold numbers of measurement errors, unidentifiable toxicological mechanisms, no understanding of the dose-response function, and the use of different models in the studies. Hence, the research does not show that $PM_{2.5}$ is the causative agent in increased mortality, so lowering its incidence might have no affect whatsoever on mortality rates. In sum, science provided little support linking $PM_{2.5}$ and mortality and therefore did not indicate, much less mandate, adoption of a new guideline for particulate matter.

CASAC confronted two substantial obstacles in responding to the question presented to it regarding an acceptable standard. The first was specific to the process in this case and involved the very short time period in which CASAC had to review the scientific literature and make recommendations. This short time period was a function of the fact that EPA had to act pursuant to a court order and the district court had demanded a response by June 1997. Of course, the district court's impatience was a product of EPA's having sat on its hands since its last review of PM in 1979. In any case, the accelerated schedule meant that CASAC had little time to study the existing literature and no opportunity to seek research that might fill in the many gaps it found in the literature. By coupling the ozone regulations to PM, EPA exacerbated the problem and nearly guaranteed that the regulations would be based on incomplete knowledge.

Although worthy of condemnation, this accelerated review was largely specific to this particular case and is not likely to recur often. A more deeply disturbing problem, however, concerns the inherent policy component of setting any standard—a problem that is endemic to the process and recurs in every case. In fact, this would be the principal reason a federal appellate court would later cite for sustaining a challenge to the regulations. EPA had provided CASAC with no decision rule by which it might recommend a specific level. As we have seen time and time again, science simply does not come packaged with a user's manual. Even assuming that the research was compelling that a linear relationship exists so that finer particles produce greater health risks, science cannot establish where the line should be drawn. At some point, necessarily, the costs to clean the air will exceed the benefits of cleaner air. But science cannot put a value on life, good health, lost jobs, or any of the other factors that must be evaluated.

CASAC, to its credit, expressly recognized the inherent limitations in its scientific-advisory role. CASAC concluded that "there is no 'bright line' which distinguishes any of the proposed standards . . . as being significantly more protective of public health." Hence, they observed, "the selection of a specific level . . . is a policy judgment." George Wolff summed up the inherent limitation of his committee's role: "This means that the [decision] to select a given level within their proposed ranges cannot be based on science."[49]

The specific standard chosen by EPA, therefore, is a policy judgment that we might expect would be based on a close assessment of the scientific research available together with a frank evaluation of the costs and benefits of one level versus another. Like the judgment envisioned for NASA in responding to a killer meteor, the decision whether to spend money on new standards must involve a balance between the real costs associated with the proposal as against the anticipated health benefits and the likelihood of successfully realizing those benefits. However inexact such a cost-benefit analysis might be, it would seem to be at the core of all legislative and administrative policy decisions. It is rather surprising to find, therefore, that the Clean Air Act actually prohibits EPA from doing this sort of cost-benefit analysis. Indeed, it is so shocking that it is impossible to take it too seriously. Taken literally, it would mean that the cleanest air technologically possible would be mandated no matter the cost. If costs or consequences were not taken into account, EPA would have to prohibit cars from the roads, due to the health hazards of carbon monoxide. Although this might alleviate the problems associated with sport utility vehicles, it is not likely to occur any time soon. If regulations were enacted wholly regardless of economic consequences, the United States economy would quickly come crashing down. Despite Congress's admonition to consider only health benefits, no one—not even Congress—really believes that EPA promulgates regulations without considering the real-world consequences of the action. In practice, therefore, despite the legal technicality limiting EPA to promulgating regulations solely to promote health, costs are an integral part of the policy-making process at EPA.

The Clean Air Act's mandate to protect health at any cost is modified in two important ways, both of which permit a surfeit of policy to shape the science. First, the Clean Air Act does not anticipate that the EPA will enact regulations that contemplate zero risks associated with any particular pollutant. Hence, some line must be identified above zero risk, the location of which must be dictated by policy. The act defines this location imprecisely, requiring no more than "an adequate margin of safety" to ensure the public health.[50] This means, in practice, that EPA selects "levels below those at which human health effects can be said to occur with reasonable scientific certainty."[51] This conservative standard is thus a policy choice that itself must be based on some scientific baseline. The idea is to identify a level at which the substance poses a serious health risk and then select a standard that is well below this risk point. But in the absence of a scientific baseline, policy alone drives the choice.

The second point at which the policy of economic consequences plays a role is less subtle, though still a deft sleight of hand. Although, as required by the Clean Air Act, EPA claims to be guided by health alone in setting air quality standards, it explicitly relies on extensive cost-benefit analyses in *implementing* new regulations. Indeed, however anomalous, it is required to do so by law. Thus, in promulgating new clean air rules, EPA must consider only benefits when setting the standard, with costs relegated to how these rules are implemented. This, it turns out, is analogous to the age-old Christian theological paradox of how God could be one while at the same time comprising the father, the son, and the holy ghost. Unlike Thomas Aquinas, however, Carol Browner preferred to simply ignore the paradox.

Not surprisingly, given the substantial costs to the states and industry associated with implementing the new PM and ozone standards, opposition was fierce. House and Senate committees held hearings, sometimes jointly, in which the various principals were called out on the carpet to testify. Bills were introduced in both houses of Congress proposing to overturn EPA's rule making. Indeed, several members of Congress even held a news conference to announce their opposition to "Carol Browner [and] her ill-conceived suggestions on lowering of the ambient air standards in both particulate matter and ozone."[52] Much of the criticism was directed at EPA's strong advocacy of exacting new standards in light of the inexact science available to support them. In addition, many legislators criticized EPA's seeming lack of appreciation for the economic costs associated with the proposed reforms.

Given that the selection of the new PM and ozone standards were the product of political preferences, it should come as no surprise that the ultimate outcome was a political compromise. Carol Browner and EPA remained steadfast on the levels chosen for these pollutants. However, some of the specifics of the implementation schedule of these new standards were put off for up to ten years. This delay had three consequences. It permitted EPA to declare a victory for clean air even though little will change in the immediate future. It permitted industry and the states breathing room to anticipate actual implementation of the new standards. And it solved the scientific uncertainty difficulty because EPA and CASAC will have an opportunity to revisit the standards in the next five-year review period before they become enforceable. The full political character of the deal struck in this context is captured in the following passage from a memorandum from President Bill Clinton to EPA setting forth his understanding of the new EPA guidelines:

> Because the EPA is establishing standards for a new indicator for PM (i.e., $PM_{2.5}$), it is critical to develop the best information possible before attainment and nonattainment designation decisions are made. Three calendar years of Federal reference method monitoring data will be used to determine whether areas meet or do not meet the $PM_{2.5}$ standards. Three years of data will be available from the earliest monitors in the spring of 2001, and 3 years of data will be available from all monitors in 2004. Following this monitoring schedule and allowing time for data analysis, Governors and the EPA will not be able to make the first determinations as to which areas should be designated nonattainment until at least 2002, 5 years from now. The Clean Air Act, however, requires that the EPA make designation determinations (i.e., attainment, nonattainment, or unclassifiable) within 2 to 3 years of revising a NAAQS. To fulfill this requirement, in 1999 the EPA will issue "unclassifiable" designation for $PM_{2.5}$. These designations will not trigger the planning or control requirements of . . . the Act.
>
> When the EPA designates $PM_{2.5}$ nonattainment areas pursuant to the Governors' recommendations beginning in 2002, areas will be allowed 3 years to develop and submit to the EPA pollution control plans showing how they will meet the new standards. Areas will then have up to 10 years from their redesignation to nonattainment to attain the $PM_{2.5}$ standards with the possibility of two 1-year extensions.[53]

A state, therefore, effectively has until 2017 to come into compliance with the new $PM_{2.5}$ standards. We thus have new guidelines whose bite has been delayed for a substantial time and, if scientific research so indicates in five years, possibly forever.

Once EPA adopted the new regulations and the political battle was largely lost, opponents did what they often do after losing in the legislative and executive branches. They sued. The judiciary, at least to date, has proved to be far more receptive to opponents' claims and far more condemning of EPA's adroit fashioning of science policy.

The legal battlefield has moved from Arizona to Washington, D.C. The Federal Circuit Court of Appeals in the District of Columbia, given its location, hears a large number of cases involving administrative agencies, and their pronouncements on the subject are usually considered the most authoritative short of those of the Supreme Court. On the whole, courts take a laissez faire approach toward agency rule making. Long ago the Supreme Court gave up trying to police Congress's allocation of power and discretion to federal agencies. Back in the dark ages of constitutional law, in the 1930s, the Court had attempted to cabin this legislative giveaway of power under what is known as the "nondelegation" doctrine. The principle behind the doctrine is straightforward: Congress has the constitutional responsibility to *make* the law, while agencies, as part of the executive department, are expected merely to "take care" that the laws are faithfully *executed*. Hence, pursuant to the nondelegation doctrine, Congress is expected to make the policy judgments that are then implemented by the agencies. But in practice this would be difficult, if not impossible, to do. Faithful application of the nondelegation doctrine would paralyze the federal government, and agencies would very quickly be put out of business. Although this result might appeal to some, the Supreme Court decided long ago that agencies are at least a necessary evil. Since the mid 1930s, the Court has not invalidated any agency action using the nondelegation doctrine.

In *American Trucking Associations, Inc.* v. *United States Environmental Protection Agency*,[54] however, the D.C. Circuit held that "the construction of the Clean Air Act on which EPA relied in promulgating the NAAQS . . . effects an unconstitutional delegation of legislative power."[55] The court found that EPA had failed to articulate any principle by which it chose the levels it designated in the regulations. The court explained, "Here it is as though Congress commanded EPA to select 'big guys,' and EPA announced that it would evaluate candidates based on height and weight, but revealed no cut-off point. . . . The reasonable person responds, 'How tall? How heavy?'"[56] By failing to identify a "determinate criterion for drawing lines," EPA had assumed unto itself too much discretion over establishing policy and, in this way, had interpreted the Clean Air Act in a way, according to the court, that violated the nondelegation doctrine.

Since the EPA considers ground ozone and PM as "nonthreshold pollutants"—that is, "ones that have some possibility of some adverse health impact (however slight) at any exposure level above zero"[57]—identifying threshold levels that would provide "an adequate margin of safety" cannot be done without either bringing the American economy crashing down or setting policy regarding the quantum of deleterious health effects that will be tolerated. Presumably, crashing the American economy is not an option. Therefore, EPA must set policy. In fact, the court in *American Trucking* grudgingly accepted that EPA had to make policy when drafting regulations, but the court insisted that it had to identify the "principles" by which those decisions were made. In ordinary human affairs, these sorts of judgments would be the product of cost-benefit analyses. But since the Clean Air Act prohibits EPA from doing this sort of balancing,[58] the court set for the EPA an impossible task. They must identify a threshold providing "an adequate margin of safety," but they may not use the only tool that can accomplish this task. The EPA must feel a lot like Alice in Wonderland:

> "There's no use trying," she said, "one *can't* believe impossible things."
>
> "I daresay, you haven't much practice," said the Queen. "When I was your age, I always did it for half-an-hour a day. Why, sometimes I've believed as many as six impossible things before breakfast."[59]

It is difficult for me to criticize EPA's rule making in the PM/ozone drama. It was a rough political compromise struck in light of inadequate scientific knowledge and with directions to act further in the future. Nonetheless, a few observations are in order. To begin with, it is not clear that the American Lung Association or the Arizona district court got what they expected. The American Lung Association, in particular, might consider the outcome a Pyrrhic victory. For the court, the scientific review process could not have been especially satisfying, however formally it adhered to the letter of the law. But the court perhaps learned the lesson of the complexity of digesting complex scientific information. The rigid time schedule imposed by the court was wholly unrealistic and EPA maneuvered easily enough to avoid it. Scientific review of the new PM standard will go forward, but it will do so on the agency's time schedule. Possibly this is all the court ever really wanted.

The only real loser in the PM/ozone drama was candor. The entire debate seemed to take place on a scale that had little to do with the actual decision. The debate was phrased almost entirely in terms of science when the science played a decidedly minor role in the actual decision. This was the reality that the D.C. Circuit had such a hard time swallowing. Although perhaps it is too much to ask, policy makers should be more forthcoming about the premises on which their proposals are based. This would mean a more prominent role for cost-benefit evaluations and a plain statement of the values that are driving the decision. This is my main point of disagreement with the way the Clean Air Act is implemented; the policy component of EPA's decision making must be made clearer. Science should not be used to hide what are essentially the true bases for decision. Democracy becomes an empty charade if real choices are obscured by scientific legerdemain.

The D.C. Circuit returned the matter to EPA so that it could identify a principle, other than cost-benefit analysis, that might justify the levels at which the agency's regulations were set. Tellingly, the court did not offer guidance on what such a principle might look like, other than the ill-formed suggestion that EPA could consider benefits as long as it did not factor in costs. This is silly. It would be like deciding whether to eat some amount of ice cream without contemplating the cost in calories. Making any decision this way would lead to disaster or, at least, extreme indigestion and obesity. In the end, the court stated that if "EPA concludes that there is no principle available, it can so report to the Congress, along with such rationales as it has for the levels it chose, and seek legislation ratifying its choice."[60]

The better solution is either for the Supreme Court to reverse the D.C. Circuit's decision or, failing that, for Congress to recognize what is inevitable. Federal agencies must calculate costs and benefits when implementing regulations. This should be made an explicit part of their responsibilities. Rather than have the EPA run to Congress for ratification of its policy judgments, Congress should oversee EPA and step in when it thinks the agency has exceeded its authority.

When Congress and Agencies Meet Science

There are certain lessons that might be drawn from the trenches in which science and policy making meet, with the goal of suggesting ways to channel and control the inevitable political dimensions of the law-making process and to make explicit what is known and what is not known as a matter of "scientific fact." Because policy and science interact so fully at the juncture of legislative and administrative decision making, such subsequent recommendations would apply to both the value choices and the factual components of the problem.

The key to an institution's effective use of science is that it be fully informed. In theory, agencies execute the will of Congress; and science is merely one tool necessary to that task. In practice, however, tension exists between these two branches. This tension is an inherent component of the checks and balances that are at the core of the American form of government. Hence, both Congress and individual agencies must have independent access to scientific information.

Most federal agencies that regularly rely on science already have ready access to scientific information, through scientific advisory committees or other boards of technical advisors. Problems arise in the administrative context when those boards are given too little time to complete the task or so little authority that they are easily ignored. This is not to suggest that science panels should be given full regulatory authority. Quite the contrary should be true. Administrators are charged with the responsibility of regulatory rule making and thus the buck stops with them. But science panels should be given clear guidelines regarding the decision to be made and should be asked for both their scientific judgment and regulatory recommendation. The respective agency administrator should be deferential to the scientific assessments of the panel but should evaluate freshly the regulatory recommendation. Both the panels' and the administrators' reasons for decisions should have to be in writing and published for public inspection.

Congress should reinstitute the OTA or establish a comparable research institution. Without its own source of scientific information, Congress is doomed to rely on agency experts and lobbyist-expert-flunkies. This is no way to run a government. Undoubtedly, a reconstituted OTA would not guarantee scientifically informed policy making any more than the old OTA guaranteed it. But a new OTA would at least provide a resource legislators could turn to when considering their own policy choices as well as choices made by the executive branch. Finally, though the suggestion is perhaps hopelessly naive, Congress should make available in written form its reasons for acting. Congress does not have to get the close review of science correct, but it should have to do a close review of science.

No Crystal Balls, Please

What the Future Holds
for Science in the Law

And thence we came forth, to see again the stars.
— Dante

"Would you tell me, please, which way I ought to go from here?" "That depends a good deal on where you want to get to", said the Cat. "I don't much care where . . . ," said Alice. "Then it doesn't matter which way you go," said the Cat. "So long as I get somewhere," Alice added as an explanation. "Oh, you're sure to do that," said the Cat, "if you only walk long enough."

— Lewis Carroll

Two themes recur throughout the law and science relationship. The first flows out of their different natures and the second is a function of the structure of the American legal process. The first concerns the essentially different philosophical orientations of the two disciplines. Science is limited to the relatively modest task of describing the "real world" or what the law refers to simply as "the facts." Unfortunately, science rarely, if ever, describes the factual world definitively. This is especially so for the sort of facts the law usually deems relevant. Identifying human carcinogens or teratogens, predicting violence, anticipating the effects of global warming, determining whether there is global warming, comparing striations on bullets, describing the effects of deforestation, and the surfeit of other factual questions the law asks are so complex that even the best research often offers only a glimpse into the underlying realities. The law, on the other hand, is largely an engine of normativity. It uses information about the world as the starting point from which it generates rules and goals for controlling the world. It must integrate the uncertain knowledge about empirical reality into a complex web of normative values held with more or less certain conviction.

The second recurring theme concerns the dynamic by which the several divisions of the law in the American system—judicial, legislative, and executive—integrate empirical data into their respective normative convictions. Each component contains in unequal parts a mixture of participatory democracy and government by experts. Judges share adjudicative responsibilities with juries, legislators are beholden to voters and influenced by lobbyists, and administrators hold public hearings and are responsible to executive and legislative overseers. For law to use science effectively, it must maneuver science through the many-faceted layers of the system and allocate responsibility for resolving legally relevant scientific disputes in accordance with underlying principles of institutional and democratic competence. The law and science intersection, then, actually comes down to a fairly specific tactical problem: allocating responsibility between legal experts (judge, legislator, and bureaucrat) and popular will (jurors, voters, and public sentiment generally) in a way that maximizes the use of science but effectuates basic democratic and constitutional principles. It is specific, but it is not simple.

The best way to solve as complex a tactical problem as integrating science into the law might be to generate a set of guidelines by which the problem can be approached. I offer the following suggestions.

1. Science (and scientists) cannot prescribe policy.
2. The error rate associated with scientific research is itself policy.
3. To make science policy, legislators, administrators, and judges must understand the science.
4. Under the United States Constitution, legislators, administrators, and judges are responsible for setting science policy.

Science (and Scientists) Cannot Prescribe Policy

Much of the tension between law and science is attributable to the competition over the "ought," that is, over value definition. Science is ostensibly positivistic in that it is limited to describing the "real world" or the world of our experience. But science is imbued with value choices and normative preferences. Some of the normative aspects of science are inherent to the industry. What questions are studied, what methods are selected to study them, what decision rules are applied to determine when a question has been answered—these are all value-driven issues that make up an essential part of the scientific enterprise. The scientific method itself is largely about limiting the individual preferences of scientists. Replication of results, use of different research paradigms to study similar questions, double-blind experimental techniques, and the convention of a 95-percent level of confidence are all designed to limit scientists' individual preferences. The model of good science is that of the detached and neutral researcher interested only in finding the "truth." While this is something of a caricature, when viewed as a whole, the scientific method has been remarkably successful at identifying truths that transcend political and cultural boundaries. Science is hyper-self-critical, even if individual scientists cannot always be trusted to be. Karl Popper observed: "The history of science, like the history of all human ideas, is a history of irresponsible dreams, of obstinacy, and of error. But science is one of the very few human activities—perhaps the only one—in which the errors are systematically criticized and fairly often, in time, corrected."[1]

Individual scientists, however, are human, and it would be remarkable if they did not have some opinions about the significance of their work for society. The twentieth century was filled with famous examples of scientists stepping into the world of policy. J. Robert Oppenheimer offered advice on nuclear policy. Edward Teller advised presidents to expand military spending and technical expertise. Ian Wilmut condemned the concept of human cloning. Albert Einstein became a statesman for peace. At some level we should welcome the interest displayed by scientists to be politically involved and to assist policy makers in dealing with cutting-edge technical issues—at least as long as they only assist the policy makers and do not seek to displace them.

The problem is specifying exactly where science, especially indeterminate applied science, ends and policy begins. Is it possible for scientists to separate their science from their politics? Consider, for example, the eminent biologist Edward O. Wilson, who is also a renowned popular science writer (and winner of two Pulitzer prizes) and a strong advocate for the environment. In June 1994, Professor Wilson testified before the Senate Subcommittee on Clean Water, Fisheries and Wildlife regarding reauthorization of the Endangered Species Act. As Senator Bob Graham explained in his introductory remarks, Professor Wilson and several other scientists would testify for the purpose of giving the subcommittee "perspective" so that it could "understand why the Act was deemed important when it was enacted 20 years ago and what we have learned since."[2] He introduced Professor Wilson by observing that "his statement, though brief, is filled with information vital to our inquiry." He added, "*We hope to learn the science of the Act from him today.*[3]

Let me preface my comments here by saying that I have the utmost respect for Professor Wilson as a scientist and science writer. I also agree politically with his very strong views on the environment. My library contains every book he has ever written, and I would vote for him for president—or at least support him for Secretary of the Interior. Indeed, it is exactly because of his very fine reputation that I have selected him as my example.

In his statement, Professor Wilson provided substantial scientifically based information. For example, he explained that "during the past 100 years total extinction has come to at least 2.3% of the bird species, 2.2% of the amphibians, 1.1% of the 20,000 plant species, and a truly alarming 8.6% of the freshwater bivalve mollusk species."[4] He noted that the current extinction rate was "between one thousand and ten thousand" times greater than "before the origin of humanity."[5] Although these statements contain a good deal of scientific information, they also are couched deeply in a particular normative view. Certainly, all things being equal, extinction is a bad thing (though this, too, is still a normative statement). But all things are rarely equal. And Professor Wilson also offered his opinion on whether the act was politically worth the costs: "[The Act] has protected over 700 threatened species of plants and animals at a proportionately modest cost to the public: of 34,000 cases reviewed under the Act from 1987 to 1991, to take one recent period, steps were taken to block only 18 developmental projects affected. In most cases, new arrangements were made without serious loss of jobs."[6]

Ironically, the Endangered Species Act does not even call for the sort of cost-benefit evaluation Professor Wilson undertook to defend it. Senator Bob Packwood (Oregon), an opponent of the act, had cosponsored an unsuccessful attempt to amend the act in 1994 to require just this sort of formal cost-benefit calculation before the act would be applied in any specific case. According to Packwood, the costs were usually not worth it: "In the Northwest alone, since the spotted owl was listed as threatened in 1990, millions of acres of federal timberland and thousands of private acres have been set aside for owls. Estimates of the number of jobs that will be lost as a result of this action range anywhere from 35,000 to 150,000."[7] Given these costs and using Professor Wilson's own science, Senator Packwood argued that the benefits did not add up: "According to Edward O. Wilson . . . there may be something on the order of 100 million species, of which only 1.4 million have been named. How many billions of dollars are we willing to spend attempting to save fungi, insects and bacteria we've never heard of and for which there may be little or no chance of recovery in any case."[8]

Professor Wilson is hardly unique in describing his science in a way that might help effectuate his politics. In fact, he was somewhat unusual in being candid that his testimony contained a large proportion of value judgments. Most scientists would prefer to keep their politics hidden behind their data. Still, scientists who advocate policy, either alongside their science or even embedded in their science, cannot be blamed entirely. The law should assume that this is an inevitable result of bringing scientists into the policy realm. And it applies to scientists of all political persuasions. One of the reasons most scientists spend innumerable hours collecting data—whether it involves crawling around the forests of the Congo collecting ants or interviewing college sophomores in Charlottesville, Virginia—is because they feel deeply about their subjects. It would be completely unrealistic to expect scientists to remain reticent on what they think "should be done" with their data when they get to testify or consult with legislators, judges, or administrators. For many scientists, this is a highlight of their careers, and they usually are not shy in telling policy makers what to do.

Of course, it would be preferable if scientists either left their value preferences at the door before going in to testify or made clear throughout their remarks when their statements were based on such preferences. It is unrealistic to expect scientists to police themselves fully. Any such policy would be the responsibility of the various professional organizations to which scientists belong. But given the number and range in perspectives of these groups, the law cannot depend on them to do the task. Finally, to some extent, scientists testifying in courts or hearings might not even be aware of how substantially normative or political considerations might have affected their conclusions. This could be due to their getting caught up in the adversarial nature of these proceedings or simply having their unconscious biases affect their judgments.

Recognizing that scientists inevitably bring their political baggage along with them when they visit the courtroom or hearing room does not mean that the law should allow them to rummage through that baggage during their testimony. Scientists, no less than other citizens, have a right to their opinions. But they do not have more right to them. Given these realities, policy makers have an obligation to critically assess the science that comes to them. Whether it is the battered woman syndrome in the trial context or ambient air quality standards at EPA or biodiversity before a Senate subcommittee, policy makers must parse the science. Only if they understand the science can they distinguish fact from advocacy. Even then it won't be easy. But with no such understanding, lawmakers are sitting ducks for the latest snake oil syndrome. The most effective check on scientists' tendency to veer into policy is to have policy makers who know enough science to keep the scientists somewhat honest.

Scientific Error Rates Are Legal Policy

In addition to political legerdemain among scientists, both conscious and unconscious, and the inevitable value choices that are endemic to doing science, virtually all the scientific findings lawmakers consider are subject to error. Whatever the hypothesis or theory and no matter how well researched, there is a possibility that the scientific conclusions are wrong. Hence, saccharin might not cause bladder cancer, sport utility vehicles might not contribute to more deaths, the 2.5 standard for particulate matter might not save lives, and rape trauma syndrome might not be diagnostic of whether a woman consented to sexual relations. Moreover, even if these hypotheses are generally correct, they might not be correct enough to support particular legal outcomes. Hence, saccharin might contribute to bladder cancer deaths but not so many that it is worth the costs imposed on users when there is no other sugar substitute. Similarly, the $PM_{2.5}$ standard might save fewer lives than expected and cost billions of dollars more than originally anticipated to implement.

The question of what benefits are worth the costs is basically a matter of allocating the burden of proof. For instance, when the first warning signs were noticed about silicone implants, the FDA ordered them off the market because the manufacturers could not demonstrate that they were safe. At that time the research was equivocal and the manufacturers had the burden of proof under applicable law. But in court, the plaintiffs had the burden of proof, and they struggled with the equivocal research available. Who must bear the burden of proof and how great the burden should be is a policy decision. It is the responsibility of policy makers, who are generally accountable in our constitutional system, to make these decisions. Also, in most legal contexts (with the primary exception of trial courts), the policy maker has the duty to say whether the burden of proof has been met. Obviously, if they cannot distinguish a mean from a median or if they believe that multiple regression analysis is something hypnotists do, then the policy judgments will not be very well informed. Alternatively, responsibility for these decisions will be abdicated to scientists or their pseudobrethren.

The trial process, in which jurors play a large fact-finding role, presents somewhat unique circumstances. As I discussed in Chapter III, the United States Supreme Court has described trial court judges as gatekeepers who are obligated to screen experts to ensure that at least a minimum amount of validity is associated with their testimony. This role is consistent with the judge's lawmaking function of determining whether the research is adequate to permit the jury to decide the matter. But judges cannot decide validity in a vacuum. They must also ask what the costs are of making an error in this and similar cases. Virtually all evidence codes, federal and state, give judges the discretion to exclude evidence if its probative value is outweighed by substantial prejudicial effect. This rule gives trial courts the power to exclude scientific evidence that is more likely than not valid but not likely enough.

Consider the example of polygraphs. Research indicates that they are probably better than flipping a coin in assessing the veracity of a witness. The general admissibility standard queries whether the basis for scientific evidence is more likely than not reliable. By this standard, polygraphs would appear to pass muster. Yet courts have been reluctant to move in this direction. And for good reason.

Polygraphs are problematic for an assortment of reasons. Polygraph practices are not well standardized, test takers vary widely in their abilities to "beat the test," and polygraphs partly displace the function of the jury to assess the credibility of witnesses. There are many reasons not to like polygraphs, even if they meet minimum standards of validity and reliability. Other forms of scientific evidence similarly present complex policy visages that require judges to look more deeply than whether the technology is better than flipping a coin. In mass toxic tort cases, such as agent orange, silicone implants, and asbestos, the huge costs to the litigants, and the costs to society of bankrupting defendants, must be taken into account. With rape trauma syndrome, the consequences for rape victims of eviscerating the rape shield statute should be considered. With forensic evidence, such as handwriting, ballistics, or bite marks, the integrity of the criminal justice system is implicated as well as the ability of defendants to meet this evidence.

This general analysis should not by any means always result in the exclusion of expert evidence. Consider the case of Jessie Ulmer and Robert Savoie.[9] Jessie and Robert's house was destroyed by fire on April 21, 1994. Soon after, they filed a claim with their insurer, State Farm. State Farm denied the claim, asserting that Jessie and Robert either had themselves or had someone on their behalf deliberately set fire to their house. As a result of these arson allegations, the state fire marshal launched an investigation. Pursuant to this investigation, the marshal asked Jessie and Robert to submit to a polygraph examination. They readily agreed. They "passed" the test, and based on these results and other information developed during the course of the investigation, the marshal concluded that they had not been behind the blaze.

State Farm, however, continued to deny their claim. Jessie and Robert sued. At trial they sought to introduce the results of the polygraph, and State Farm objected on the basis that polygraph tests were insufficiently reliable. Judge Little, in an opinion applying the new *Daubert* standard, admitted the polygraph results. Despite my general skepticism about polygraphs and my agreement with the many cases since *Daubert* that have excluded polygraph testimony, Judge Little made the right decision here. And he did it for the right reasons. Judge Little began by noting that the polygraph has been studied intensively and is considered by many to produce results that exceed chance.[10] In addition, the general value of polygraphs was buttressed in this case by the circumstances surrounding the administration of the test. Although State Farm had not participated in the test, they had instigated the investigation of which the test was a part. They could not now complain when the investigation led to an unwelcome result. Moreover, the polygraph administrator was neutral at worst, and Jessie and Robert had put themselves in jeopardy by agreeing to be questioned. Finally, there was a genuine need for the evidence because it substantiated Jessie and Robert's claim not to have been involved in setting the fire.

Without question, there is a significant danger associated with calling on trial courts to conduct a sophisticated evaluation of scientific evidence in light of the costs presented by various legal contexts. It could become a gaping loophole into which a surplus of bad science might surge. But I think this is a danger worth the risk. Foremost, this is a judge's job. The danger that some judges will fail to do their duty is simply not a reason to excuse them from this duty. In addition, judges have the capacity to deal with complex scientific information. Many legal ideas rival scientific concepts in difficulty. The problem lies in convincing judges of the need to invest the time in learning the science. But I think that time has come.

Legislators, Administrators, and Judges Must Understand the Science

The natural question asked by many, including many judges, legislators, and administrators, concerns just how lawyers are to learn enough science to supervise the scientists effectively. This is a fair question. Until this point, my attention has focused on the need for lawmakers to oversee the science. The biggest problem confronting policy makers is their need to recognize that they have a problem. There is little hope for any treatment if the policy makers believe that the status quo is satisfactory. Much like overcoming drugs, alcohol, or cigarettes, the first step to good health is recognizing the need for help. Learning the science will be difficult for many lawmakers, and it will take some motivation. Hence, the first task is convincing them that they must invest the time. Assuming for the moment that I have done that, I will now turn to what I consider the subsidiary question of how to educate lawmakers on the rigors of the scientific method and the niceties of scientific culture.

Most self-help programs seem to be divided into twelve steps to make the ultimate end appear more manageable and attainable. Although meant to be facetious, this model does serve my purposes rather well. As with most self-help programs of the sort needed for policy makers, success can occur only after the individual has completed all twelve steps. The program could be called "Innumerates Anonymous" and be open to anyone with a law degree or a policy-making job in local, state, or federal government. The twelve steps must be followed in order for best results.

1. I am an innumerate.
2. The law needs the best science available or that could be made available.
3. As a/an _____ (insert one: practicing attorney, professor, judge, administrator, legislator, or other), it behooves me to become familiar with science and the scientific culture in order to fulfill my professional obligations.
4. I don't have to be a scientist to understand the rudiments of science.
5. I have the ability and motivation to learn about science.
6. Science is deeply fascinating and genuinely fun.
7. I am not afraid to seek help from bona fide experts in relevant fields, but I will not abdicate my responsibility to them.
8. I am willing to attend continuing education programs to help me, and if necessary and when relevant and useful, I will read published reference materials on science.
9. I will not label something "science" when it is merely the product of advocacy and has not been tested.
10. I will endeavor to understand the nuts and bolts of the scientific method and not simply the conclusory testaments offered by scientists or those pretending to that title.
11. I will call on all experts to provide the best information that their methods or methods that could be employed can deliver.
12. I am now a sophisticated consumer of science.

The philosophy behind this twelve-step program is to reach the twelfth step when the legally trained generalist feels prepared for the task of intelligently integrating scientific knowledge into legal decision making. I have to emphasize, however, that my goal is not to turn lawyers into scientists. They must merely be good consumers of science. The difference is significant. It is the difference between being able to build a car and being able to evaluate a car's performance and track record before buying it. Lawyers need not have the ability to conduct research and write a report of it. But they should have the ability to read research reports written by scientists.

For the most part I am not talking about rocket science here. In the courtroom, most of the science that presents such debatable questions suffers from relatively stark methodological problems. Indeed, in many cases, such as bite marks, tool marks, handwriting, fiber analysis, and the sundry syndromes popping up regularly, there is virtually no research at all. In the case of the battered woman syndrome, for instance, the methodological flaws are so plain that a bright college junior majoring in psychology should be able to identify most of them. It is not the twentieth-century physics of Albert Einstein that judges are having difficulty with; it is the sixteenth-century inductive reasoning of Francis Bacon that they don't seem to understand.

Outside the courtroom, the science sometimes can be more difficult to understand. But even when policy makers are evaluating "rocket science," which they sometimes must do, they typically do not need a deep understanding of the physics involved. The value of the supercollider or the human genome project can be assessed without knowing the math behind Higgs particles or the chemical composition of the four DNA nucleotide bases. Some of these subjects might take more time and be more challenging than studying, say, the literature behind bite mark expertise. But as lawyers increasingly recognize the relevance of science and the need to learn it, the market will supply plenty of quality assistance to help them.

Every branch of government is familiar with receiving technical assistance from one quarter or another. Although Congress killed it in 1995, the Office of Technology Assessment provided detailed reports to assist legislators understand the premises behind scientific and technical conclusions. Today, legislators receive technical information on occasion from the Congressional Research Service, the General Accounting Office, the National Research Council, the National Academy of Sciences, and more frequently though less neutrally, lobbyists, corporations, and interested citizens. Unfortunately, none of these sources provides the depth of scientific analysis once provided by OTA. I reiterate my conviction that Congress should seriously reconsider its dreadful decision to abolish OTA. The judiciary, especially since *Daubert*, also employs experts to provide technical assistance separate from the parties' experts. This primarily takes the form of special masters or court-appointed experts whose only allegiance is to the court (and presumably to the "truth," as well as it can be determined). In addition, courts more recently, especially in the mass toxic tort area, have begun experimenting with panels of experts selected from different fields. In the silicone implant litigation alone there were two such panels, one serving Judge Robert Jones in Portland, Oregon, and the other assisting Judge Samuel Pointer in Montgomery, Alabama. Judge Pointer's panel, for example, consisted of four scientists, in the areas of toxicology, medicine, epidemiology, and immunology. Administrative agencies are probably most familiar with this kind of institutional technical support. All agencies that deal substantially with scientific topics have scientific advisory boards that are primarily responsible for parsing and explaining the science. In fact, courts and legislatures might fruitfully study the administrative framework of scientific advisory committees for lessons and insights for their own consultation needs.

However, court-appointed experts, scientific advisory boards, and legislative advisory councils cannot be permitted to substitute for the reasoned judgments of policy makers. The principal danger of this sort of expert assistance is that policy makers will abdicate their responsibilities to it or will not have the ability to critically assess what their advisors give them. They would thus blindly follow scientific recommendations without really appreciating the value choices inherent in the recommendations. There is just no alternative to learning the science.

The judiciary appears to have learned this lesson. Since 1993, when *Daubert* was decided, seminars, workshops, and other continuing education programs have sprung up around the country. The Federal Judicial Center, a research and support arm of the federal judiciary, has held regular programs on basic statistical methods, epidemiology, and other related subjects. The National Judicial College in Reno, Nevada, holds several programs designed to help judges with science, ranging from summer classes leading to advanced law degrees with concentrations in scientific areas to week-long seminars on scientific evidence. The Private Adjudication Center at Duke University has an annual week-long workshop on scientific evidence for state and federal judges, and many states have sponsored similar programs for their judges. Finally, there are now several reference books available specifically designed to bring the science behind scientific evidence to judges and lawyers.[11]

Understanding science is part of the constitutional duty assumed by legislators, administrators, and judges. Science policy can be made only with a good understanding of the science and a deep sense of responsibility for the policy. The lesson of example after example in this book has been that science at the policy level is about both science and policy. When scientific research—or the lack thereof—underlies the decision, legislators cannot "make law" satisfactorily, executive agents cannot "take care" that the "law be faithfully executed," and judges cannot "say what the law is" if they are ignorant of science.

Legislators, Administrators, and Judges Are Responsible for Setting Science Policy

Earlier, I made the point that law and science have common origins in religion. I return to this point here. Before the scientific revolution, the high priests had responsibility for describing the empirical world, determining the rules that controlled human behavior, and decreeing what must be done to get to the next world. Today we think of these functions as the respective realms of scientists, lawyers, and religious leaders, the high priests of modern society. This admittedly simplified view of the division of responsibility between science, law, and religion suggests a complementarity between the three disciplines. In short, as long as each institution sticks to what it's good at, they should get along just fine. But we know they do not; nor is it likely that they will any time soon.

Any proposed solution to the conflicts inherent among law, science, and religion must take into account their three overlapping natures. A solution urging simply that they stick to their respective domains would and should fail. The competition among science, law, and religion is a salutary thing. It is analogous to the checks and balances created by the three branches of the federal government. We would not want any one of these institutions to dominate the other two. History and literature are filled with examples where one of these institutions became powerful enough to displace the other two. An all-powerful science would be Huxley's *Brave New World*. Fascism, Nazism and communism showed us worlds where science and religion became servants of the state and the law. And Afghanistan and Iran illustrate religion's ascendance to power. Liberal democracy needs science, law, and religion each to be a strong presence to check the excesses of the other two.

The law must employ science well and respect the tenets of religion. Instead of the term religion, it might be more accurate to refer to schools of thought regarding the definition of morals, norms, or values. It turns out that from the law's perspective, at least as defined by the religion clauses of the First Amendment, this is what is meant by "religion." While it remains unclear whether Thoreau's philosophical views are included in the Free Exercise Clause of the Constitution, all faiths, including lack of faith, must be respected by the state. The power of religious faith for the law comes from whatever independent force it might have. The Establishment Clause prohibits the state from adhering to religious principles because they are religious principles. But the Establishment Clause certainly does not preclude the state from adopting religious prescriptions when they are supported by good secular reasons. The point is that the law must interpose its independent judgment between religious injunction and state enactment.

The law stands in a similar but not identical relationship to science. There is no constitutional amendment that proscribes the state from establishing science. (The free exercise of science, however, falls partly into the First Amendment's guarantee of free speech.) This is a good thing, too. In the religious context, government is disabled from choosing between religions or otherwise determining that one religion is more valid than another. A similar prohibition in the context of science would mean that the state could not prefer Darwin over Lysenko, the Soviet biologist who maintained the possibility of inheriting environmentally acquired characteristics. The state's task, in fact, is just the opposite. It is the obligation of policy makers to identify the best science available and, moreover, to demand that the best science be done over time.

Regarding religion, therefore, the Constitution first dictates that government can follow religious principles (or secular morality such as Kantian philosophy) if it gives secular reasons for doing so. It cannot endorse any particular set of religious beliefs, but it can enact laws that are informed by moral principles with deep religious roots. Admittedly, this is a high-wire act that has resulted in many thrills and spills, all recounted in the many volumes of First Amendment litigation. American constitutional democracy thus protects the individual's right to his or her faith but expects government to offer something more tangible.

Science is a substantial part of the equation. The Constitution imparts an affirmative obligation on lawmakers to act rationally in carrying out their public duties. This means necessarily that they should have some facility with science and the scientific method. Twenty-first-century lawmakers without a full appreciation of science will be in dereliction of their constitutional duties. We have already visited many examples of such dereliction. Justice Brennan's opinion in *Craig* v. *Boren* striking down an Oklahoma law for having an insufficient empirical basis, accompanied by his admission that he did not understand the statistics on which it was based, is one example. Judge McKenna in *United States* v. *Starzecpyzel* offers another in his allowance of handwriting identification expertise based on the unchecked experience of the witness for the prosecution. We have seen a multitude of additional examples of the constitutional necessity of having scientifically literate leaders in Congress and administrative agencies.

Science policy is an organism that cannot be separated into its constituent parts. To be sure, science and policy have separate existences that can each be studied and pursued in their own right. But science policy is not simply a combination of science on the one hand and policy on the other. When combined, they constitute a new creature altogether. At the risk of mixing my mythological metaphors, science policy is like the mythological Minotaur, which was half human and half bull. It was neither human nor bull. And it is worth recalling that the Minotaur resided in the famous Labyrinth, from which it was impossible to escape. Much of the story of the law's use of science resembles the flight of the innocent youths who were sacrificed to the Minotaur as they ran along the endlessly twisting paths of the Labyrinth in futile search of an exit. Only Theseus was successful in escaping the Labyrinth; after slaying the Minotaur, he retraced his steps by following a ball of thread which he had fastened to the door and unwound as he walked on.

Lawmakers similarly must be given a ball of thread so that they can maneuver through the labyrinth. Returning to my first metaphor, the hydras, the law has done an admirable job of borrowing valuable insights from religion and moral philosophy for its own purposes without being overwhelmed or overly influenced by them. The law largely understands the dangers associated with the religion hydra, as well as what it has to offer policy formation. There are exceptions, but the law has done well in identifying worthy ideas and avoiding ones that would lead it astray. This is largely due to the fact that lawyers understand the language of normative discourse. It is a language pervasive in most lawyers' training and is a central component of the language of law. While, to be sure, most law professors do not dwell on the eternal consequences of particular conduct, notions such as justice, fairness, and equality are daily subjects of law school classrooms.

The science hydra, however, remains a mystery to the law. The law has proved largely unsuccessful in integrating the lessons of the scientific method into its decision making. This by no means requires the law's loss of its autonomy to the sciences. Quite the contrary is true. The law cannot simply defer to what scientists and pretenders to science generally believe. It is only by understanding and sharing a part of the scientific culture that law can be a wise and effective borrower of the scientific product.

In the end, most lawyers and lawmakers are unlikely to ever aspire or achieve status as "amateur scientists." They have neither the desire nor the need to do so. On the other hand, should they become good consumers of science, the rewards will be substantial. Plato, through the mouth of Socrates, described it thus: "It seemed to me a superlative thing to know the explanation of everything, why it comes to be, why it perishes, why it is."[12]

The law will never become a sophisticated consumer of science until the lawyers and lawmakers become conversant in the language of science and are comfortable in its culture. But science does not exist like a schoolhouse that can be entered and exited as necessity requires. To be scientifically literate does not involve memorizing the terms or supposed content of "science." Science is an approach or methodology. It is a direction, not a destination. As Einstein put it, in science "imagination is more important than knowledge." It is a rigorous and critical form of analysis, one that seeks well-supported and parsimonious explanations for empirical questions.

The policy makers in the United States are ultimately responsible to "we the people." Air pollution, toxic waste, harmful drugs, forensic hocus-pocus, and endangered species are matters that are our responsibility. The United States is a noble experiment in which the people are sovereign. We have the responsibility to ensure that our elected leaders, our administrators, and our judges set science policy wisely. To fulfill this function, the obligation to understand the science in science policy cannot be limited to government officials alone. The government we deserve is the government we get. It is no longer sufficient, to be an informed citizen, to understand merely the art of politics. Forevermore, the good citizen and the good government will have to have a strong education in both the arts and sciences of policy. Although science can never dictate what is fair and just, it has become an indispensable tool on which the law must sometimes rely to do the fair and just thing.

Notes

◆

Preface

1. Mark Twain, *The Adventures of Huckelberry Finn*, 21 (1996).
2. The one law school requiring such courses is George Mason University School of Law.

Chapter 1

1. The story of Roland Molineux recounted here is based on *American Trials: The Molineux Case*, ed. Samuel Klaus (1929), and the original transcript of the trial and the opinion of the appellate court, *People v. Molineux*, 61 N.E. 286 (1901).
2. Klaus, note 1, at 218.
3. *Id.* at 237.
4. *Id.* at 239.
5. The text that remains the most authoritative guide to the forensic practice of handwriting identification was published shortly after the Molineux case: Albert S. Osborn, *Questioned Documents* (1910). This work gathered and described many of the prevailing practices of Molineux's day, thus bringing some order to this field. Before Osborn, several volumes provided similar advice to the practicing analyst. *See* Charles Chabot, *The Handwriting of Junius Professionally Examined*, ed. Edward Twistleton (1871); William E. Hagan, *Disputed Handwriting* (1894); D. T. Ames, *Ames on Forgery* (1899).
6. *See* D. Michael Risinger, *Handwriting Identification*, § 22 *in* Faigman et al., *Modern Scientific Evidence: The Law and Science of Expert Testimony* (1997).
7. Klaus, note 1, at 35.
8. *See* David C. Lindberg, *The Beginnings of Western Science: The European Scientific Tradition in Philosophical, Religious, and Institutional Context, 600 B.C. to A.D. 1450*, 2–5 (1992).

9. *See* Michael White, *Isaac Newton: The Last Sorcerer* (1997).

10. *See generally* Lindberg, note 8, at 5.

11. Riesman, *Some Observations on Law and Psychology*, 19 U. Chi. L. Rev. 30, 32 (1951); *see also* Steven Goldberg, *The Reluctant Embrace: Law and Science in America*, 75 Geo. L. J. 1341 (1987).

12. Daniel J. Kevles, *In the Name of Eugenics*, 12 (1985, 1995).

13. *Id.*

14. Kevles, note 12, at 4.

15. *Id.*

16. Kevles, note 12, at 40.

17. *Id.*

18. Kevles, note 12, at 93–94.

19. Declaration of Independence, para. 1 (U.S. 1776).

20. *See* Adrian Desmond & James Moore, *Darwin: The Life of a Tormented Evolutionist* (1991).

21. The full quote is "When I hear anyone talk of culture, I reach for my revolver." *The Penguin Dictionary of Twentieth Century Quotations*, 196, ed. J. M. Cohen and M. J. Cohen (1995).

22. Stephen W. Hawking, *A Brief History of Time: From the Big Bang to Black Holes*, 175 (1988).

23. William Blake, *Poems from The Pickering Manuscript, Auguries of Innocence*, 1 (1805).

24. Pascal, *Pensees*, 278, trans. W. F. Trotter (1670).

25. Lindberg, note 8, at 53.

26. *See* Thomas Henry Huxley, *The Method of Zadig* (1878).

27. Aristotle, *De Caelo*, Bk. II, Ch. 14, 279a4–8.

28. For financial reasons Copernicus actually received his degree from the University of Ferrara. *See* Jerome J. Langford, *Galileo, Science and the Church*, 34 (3d ed., 1995).

29. Copernicus, *On the Revolutions of the Heavenly Spheres*, Great Books of the Western World XVI, 508 (1952).

30. Martin Luther, *Tischreden* XXII, 2260, ed. Walsch.

31. *Opere*, XIX, at 306–361, *quoted in* Langford, note 28, at 150.

32. *Id.* at 361, *quoted in* Langford, note 28, at 151.

33. *Id.* at 362, *quoted in* Langford, note 28, at 151.

34. *Id.* at 405–406, *quoted in* Langford, note 28, at 152.

35. Charles Brockett & Billy Wilder, *Sunset Boulevard* (1950).

36. *Scopes v. State*, 289 S.W. 363 (Tenn. 1927).

37. *Id.* at 363, note 1.

38. 393 U.S. 97 (1968).

39. 482 U.S. 578 (1987).

40. *Edwards v. Aguillard*, 778 F.2d 225, 227 (5th Cir. 1985) (Gee, J., dissenting).

41. *Edwards*, 482 U.S. at 634 (Scalia, J., dissenting).

42. *Id.* at 621–622.

43. *Id.* at 622.

44. *Kansas v. Hendricks*, 521 U.S. 346 (1997).

45. *M'Naghten's Case*, 1843–60 All E.R. 229 (H.L. 1843); *see generally* Rollin, *Crime and Mental Disorder: Daniel M'Naghten, a Case in Point*, 50 Medico-Legal J. 102 (1982).

46. *M'Naghten's Case*, 1843–60 ALL E.R. at 233.

47. Model Penal Code § 4.01(1) (1985).

48. 18 U.S.C. § 17 (1988). The Congressional reform differed from the traditional *M'Naghten* test in several respects, including that it put the burden of proof on the defendant and required that the mental disease or defect be severe to qualify under the test.

49. Kan. Stat. Ann. § 59-29a01 *et seq.* (1994).

50. In re the Care and Treatment of Leroy Hendricks, 912 P.2d 129 (Kan. 1996).

51. 504 U.S. 71 (1992).

52. *Foucha*, 504 U.S. at 77, quoting *Jones v. United States*, 463 U.S. 354, 368 (1992).

53. *Foucha*, 504 U.S. at 76, quoting *Jones*, 463 U.S. at 363.

54. 521 U.S. 346 (1997).

55. *Id.* at 358.

56. *Id.*

57. *Id.* at 360.

58. *Id.* at 375 (Breyer, J., dissenting).

59. Kan. Stat. Ann. § 59-29a02(b)(1994).

60. *See* Eric S. Janus & Paul E. Meehl, *Assessing the Legal Standard for Predictions of Dangerousness in Sex Offender Commitment Proceedings*, 3 Psychology, Public Policy & Law 33 (1997).

61. Kan. Stat. Ann. § 59-29a02(a) (1994).

62. *Hendricks*, 521 U.S. at 357.

63. Justice Breyer made this point in reaching his conclusion that the Kansas statute was in fact criminal punishment: "The Act, like criminal punishment,

imposes its confinement (or sanction) only upon an individual who has previously committed a criminal offense." *Hendricks*, 117 S. Ct. at 2091 (Breyer, J., dissenting).

64. The Court also noted that there was no scienter requirement in the law, which is "customarily an important element in distinguishing criminal from civil statutes." *Id.*

65. For a very good general history of the witch trials, *see* Frances Hill, *A Delusion of Satan: The Full Story of the Salem Witch Trials* (1997).

66. Ralph Boas & Louise Boas, *Cotton Mather: Keeper of the Puritan Conscience*, 91 (1964).

67. *See* Hill, note 65.

68. John Gaule, *Select Cases of Conscience Touching Witches and Witchcraft*, 4–5, 53, 54 (1646), *quoted in* Peter Charles Hoffer, *The Devil's Disciples: Makers of the Salem Witchcraft Trials*, 69 (1996).

69. Richard Baxter, *The Certainty of the Worlds of Spirits* (1691), *quoted in* Hoffer, note 68, at 69.

70. Hoffer, note 69, at 120.

71. *Id.* at 180; Hill, note 65, at 166.

72. *Id.*

73. Hoffer, note 68, at 66.

74. *Id.* at 145.

75. *Id.* at 78, quoting *The Book of General Lawes and Liberties of 1648* at 5.

76. *Id.* at 149.

77. *Id.*

78. *Id.*

79. *See* Hoffer, note 68, at 191.

80. Boas & Boas, note 67, at 95.

81. Mason I. Lowance, Jr., *Increase Mather*, 103 (1974).

Chapter 2

1. 842 S.W.2d 588 (Tenn. 1992).

2. 1986 La. Acts R.S. 9:121 *et seq.*

3. Alexis de Tocqueville, *Democracy in America*, 357–358 (1835, trans. Francis Bowen, 1862).

4. 1989 WL 140495 at *3.

5. Sally Jacobs, *Woman Awarded Custody in Frozen Embryo Case*, Boston Globe, September 22, 1989.

6. George J. Annas, *A French Homunculus in a Tennessee Court,* 19 Hastings Center Rep. 20, 21 (1989).

7. 1989 WL 140495 at *3.

8. Annas, note 88, at 21.

9. 842 S.W.2d at 593.

10. Mark Curriden, *Embryos Are People, Says Expert on Genetics,* Atlanta Journal, August 11, 1989.

11. Wendy Holden, *Frozen Embryo Case a Tough One for Judge,* Vancouver Sun, August 12, 1989.

12. Annas, note 88, at 22.

13. *Davis,* 842 S.W.2d at 597 (emphasis added).

14. 497 U.S. 261 (1990).

15. *Id.* at 277–278, quoting *Twin City Bank* v. *Nebeker,* 167 U.S. 196 (1897).

16. *Id.* at 278.

17. 117 S. Ct. 2258 (1997).

18. 117 S. Ct. 2293 (1997).

19. *Glucksberg,* 117 S. Ct. at 2274.

20. Alexander Bickel, *The Least Dangerous Branch* (1962).

21. Or. Rev. Stat. T.13, ch.127 (1998).

22. 21 U.S.C. §841.

23. David Brandt-Erichsen, *Feds Trying to Block Oregon Law,* Los Angeles Times, A21, November 12, 1997.

24. *Id.*

25. 143 Cong. Rec. S12235-01, daily ed. November. 9, 1997, statement of Sen. Wyden.

26. *Id.*

27. 143 Cong. Rec. S3249-01, daily ed. April 16, 1997, statement of Sen. Ashcroft.

28. 144 Cong. Rec. S5787-03, *S5797, daily ed. June 9, 1998, statement of Sen. Nickles.

29. An excellent overview of the general interaction between law and science, though written for a more technical audience, is Professor Steven Goldberg's *Culture Clash: Law and Science in America* (1994). Professor Sheila Jasanoff has written on two of these arenas, courts and administrative agencies. *See* Sheila Jasanoff, *Science at the Bar: Law, Science and Technology in America* (1995); Sheila Jasanoff, *The Fifth Branch: Science Advisers as Policymakers* (1990). Although Professor Jasanoff's approach and perspective are different from my own, her books are scholarly and

thought provoking. The interested reader is encouraged to peruse her work. In general, trial courts have received the most attention, though many are organized around a specific issue or case. More than a dozen books were published on the O. J. Simpson case alone. Many books focus on some specific issue, such as Marcia Angell's *Science on Trial* (1996), which focused on, in her view, the courts' failure to understand the science in the silicone implant cases. One noteworthy general treatment of trial courts, because it popularized the term "junk science," is Peter Huber's *Galileo's Revenge: Junk Science in the Courtroom* (1991). It is, however, a generally myopic attack on expert witnesses, especially those who testify for plaintiffs. In the administrative agency arena, a very helpful book is Bruce L. R. Smith, *The Advisers: Scientists in the Police Process* (1992). There have been no general studies or assessments of the legislative use of science.

30. Telephone interview with Mr. Robert Carr, Director of Research, Law School Services Corp. (June 30, 1998).

31. *See, e.g.,* Steven Goldberg, *Culture Clash: Law and Science in America* (1994); Philip M. Boffey, *Scientists and Bureaucrats: A Clash of Cultures on FDA Advisory Panel*, 199 Sci. 1244 (1976); Leslie Roberts, *Science in Court: A Culture Clash*, 257 Sci. 732 (1992); Peter H. Schuck, *Multi-Culturalism Redux: Science, Law, and Politics*, 11 Yale L. & Pol. Rev. 14 (1993).

32. *See* Thomas Kuhn, *The Structure of Scientific Revolutions* (1973).

33. T. S. Eliot, *The Wasteland*, pt. 5 (1922).

Chapter 3

1. For an overview of the science and the litigation surrounding silicone-gel breast implants, *See* Marcia Angell, *Science on Trial* (1996).

2. Franz Kafka, *The Trial* 267–268 (1964).

3. *Id.*

4. The three cases are *Daubert* v. *Merrell Dow Pharmaceuticals, Inc.*, 509 U.S. 579 (1993)—Bendectin and birth defects; *General Electric Co.* v. *Joiner*, 118 S. Ct. 512 (1997)—PCBs and lung cancer; and *Kumho Tire Co.* v. *Carmichael*, 118 S. Ct. 2339 (1999)—tire engineer. A fourth case, *United States* v. *Scheffer*, 118 S. Ct. 1261 (1998), also involved scientific evidence—polygraph tests—but was decided primarily on constitutional grounds.

5. For an excellent overview of the entire history of litigation over Bendectin, *see* Joseph Sanders, *The Bendectin Litigation: A Case Study in the Life Cycle of Mass Torts*, 43 Hastings L. J. 301 (1992).

6. *Daubert v. Merrell Dow Pharmaceuticals, Inc.*, 43 F.3d 1311, 1316 (9th Cir. 1995), *cert denied*, 116 S.Ct. 189 (1995).

7. *Id.* at 1317.

8. *Id.* at 1315.

9. Since the *Smith* trial, many jurisdictions, including the federal courts, now permit evidence of other sexual assaults in order to avoid the perceived miscarriage of justice in cases like Smith's. *See* Fed. R. Evid. 413, 414, and 415.

10. For an outstanding study of the use of burdens of proof, *See* Richard H. Gaskins, *Burdens of Proof in Modern Discourse* (1992).

11. Alan M. Dershowitz, *The Abuse Excuse and Other Cop-outs, Sob Stories, and Evasions of Responsibility* (1994). For other indictments of the criminal justice system's response to syndrome-styled expert testimony, *see* James Q. Wilson, *Moral Judgment: Does the Abuse Threaten Our Legal System?* (1997); Judge Harold J. Rothwax, *Guilty: The Collapse of Criminal Justice* (1996); George P. Fletcher, *With Justice for Some: Protecting Victims' Rights in Criminal Trials* (1995).

12. *See* Ann Burgess & Linda Holmstrom, *Rape Trauma Syndrome*, 131 Am. J. Psychiatry 981 (1974).

13. Lenore Walker, *The Battered Woman* (1979).

14. Lenore Walker, *The Battered Woman Syndrome* (1984).

15. *See* David L. Faigman, Note, *The Battered Woman Syndrome and Self-Defense: A Legal and Empirical Dissent*, 72 Va. L. Rev. 619 (1986).

16. *See, e.g., State v. Martin*, 666 S.W.2d 895 (Mo. Ct. App. 1984).

17. Lord Chesterfield, *Letters to His Son*, October 4, 1746.

18. Thomas Henry Huxley, *On Elemental Instruction in Physiology* (1877).

19. 880 F.Supp. 1027 (S.D.N.Y. 1995).

20. For a fuller description of this study, *see* D. Michael Risinger, Mark P. Denbeaux, & Michael J. Saks, *Exorcism of Ignorance as a Proxy for Rational Knowledge: The Lessons of Handwriting Identification "Expertise,"* 137 U. Pa. L. Rev. 731 (1989).

21. 25 F.3d 1342 (6th Cir. 1994).

22. *Carmichael v. Samyang Tire, Inc.*, 131 F.3d 1433, 1435–36 (11th Cir. 1997).

23. *Kumho Tire Co. v. Carmichael*, 119 S. Ct. 1167 (March 23, 1999). Justice Stevens dissented from one part of the decision, since he believed the Court should have remanded the case back to the Eleventh Circuit so that that court could determine whether the trial court had abused its discretion in excluding the engineer's testimony. *Id.* at 1179 (Stevens, J., concurring in part and dissenting in part).

24. *Id.* at 1171.

25. *Id.* at 1174.

26. *Id.*

27. *Id.*

28. *Daubert*, 509 U.S. at 591.

29. *See* Stan Liebowitz & Stephen Margolis, *The Fable of the Keys*, J. L. & Econ. (April 1990).

30. *Kumho*, 119 S. Ct. at 1175.

31. *Id.*

32. *Id.* at 1176.

33. *Id.* at 1175.

Chapter 4

1. Henry Adams, 1 *History of the United States During the Administration of Thomas Jefferson*, 132 (1986).

2. William Marbury was not technically one of the "midnight judges" created under the Circuit Court Act passed on February 13, 1801. His position was created even later, in the Organic Act of the District of Columbia, passed on February 27, 1801. Adams's term ended one week after the Organic Act was passed, on March 3, 1801.

3. 5 U.S. (1 Cranch) 137 (1803).

4. *McCulloch v. Maryland*, 17 U.S. (4 Wheat) 316 (1819).

5. *Id.* at 177.

6. *Id.*

7. *Id.*

8. Thomas Jefferson, *Letter to William C. Jarvis*, September 28, 1820, in *Writings of Thomas Jefferson*, 10:160, ed. Paul L. Ford (1899).

9. 2 Farrand, *The Records of the Federal Convention of 1787*, 172 (1911).

10. *Legal Tender Cases*, 79 U.S. (12 Wall.) 457 (1871).

11. *Helvering v. Hallick*, 309 U.S. 106, 119, 121 (1940).

12. Learned Hand, *The Bill of Rights*, 73 (1958).

13. 198 U.S. 45 (1905).

14. *Id.* at 64.

15. *Id.* at 61.

16. 208 U.S. 412 (1908).

17. *Id.* at 416.

18. *Lochner*, 198 U.S. at 75 (Holmes, J., dissenting).

19. Oliver Wendell Holmes, *The Common Law* 1 (1881).

20. Oliver Wendell Holmes, *Book Notices*, 7 Am. L. Rev. 318 (1873).

21. Oliver Wendell Holmes, *Book Notices*, 6 Am. L. Rev. 132, 141 (1871).

22. 274 U.S. 200 (1927), overruled by *Skinner v. Oklahoma*, 316 U.S. 535 (1942).

23. *Id.* at 205.

24. *Id.* at 207 (citation omitted). *See* Dudziak, *Oliver Wendell Holmes as a Eugenic Reformer: Rhetoric in the Writing of Constitutional Law*, 71 Iowa L. Rev. 833 (1986).

25. Michael Perry, *The Authority of Text, Tradition, and Reason: A Theory of Constitutional "Interpretation,"* 58 S. Cal. L. Rev. 551, 552–53, note 5 (1985).

26. 347 U.S. 483 (1954).

27. *See* John Monahan & Laurens Walker, *Social Authority: Obtaining, Evaluating, and Establishing Social Science in Law*, 134 U. Pa. L. Rev. 477, 483–484 (1986).

28. *Brown*, 347 U.S. at 494.

29. *See* Monahan & Walker, note 27, at 483–484.

30. *See, e.g.,* Berger, *Desegregation, Law, and Social Science*, 23 *Commentary* 471, 476 (1957) ("We may reach a point where we shall be entitled to equality under law only when we can show that inequality has been or would be harmful.").

31. Robert Bork, *Neutral Principles and Some First Amendment Problems*, 47 Ind. L. J. 1, 13 (1971).

32. R. Kluger, *Simple Justice* (1976).

33. Doyle, *Can Social Science Data Be Used in Judicial Decisionmaking?* 6 J. L. Educ. 13, 18 (1977). "We would pose a greater danger to our 200 year experience in constitutional interpretation if we were to rest our constitutional interpretation on social science data rather than on the bedrock of a coherent constitutional principle."

34. *Brown*, 347 U.S. at 493.

35. *Id.* at 494.

36. 220 F. Supp. 667 (S.D. Ga. 1963).

37. *Id.* at 678.

38. *Stell v. Savannah-Chatham County Bd. of Educ.*, 333 F.2d 55 (5th Cir. 1964).

39. Ronald Dworkin, *Social Sciences and Constitutional Rights—The*

Consequences of Uncertainty, 6 J. L. Educ. 3, 5 (1977), quoting Cahn, *Jurisprudence,* 30 N.Y.U. L. Rev. 150, 157–58 (1955) (emphasis added).

40. *Id.*

41. *See* Ronald Dworkin, *Law as Interpretation,* 60 Tex. L. Rev. 527, 540–46 (1982).

42. R. Dworkin, *Law's Empire,* 52 (1986).

43. *Plessy v. Ferguson,* 163 U.S. 537 (1896).

44. *Id.* at 544.

45. *Id.* at 551.

46. 410 U.S. 113 (1973).

47. *Id.* at 163.

48. *Id.*

49. *Id.*

50. *Id.*

51. 462 U.S. 416 (1983).

52. *Id.* at 458 (O'Connor, J., dissenting).

53. *Akron,* 462 U.S. at 429, note 11 (emphasis added).

54. John Hart Ely, *The Wages of Crying Wolf: A Comment on* Roe *v.* Wade, 82 Yale L. J. 920, 924 (1973).

55. 505 U.S. 833 (1992).

56. *Id.* at 860.

57. *See generally* S. R. Schlesinger & J. Neese, *Justice Harry Blackmun and Empirical Jurisprudence,* 29 Am. U. L. Rev. 405 (1980).

58. 435 U.S. 223 (1978).

59. 399 U.S. 78 (1970).

60. *Id.* at 100 (emphasis added).

61. *Ballew,* 435 U.S. at 239.

62. David H. Kaye, *And Then There Were None: Statistical Reasoning in the Supreme Court, and the Size of the Jury,* 68 Cal. L. Rev. 1004, 1008 (1980).

63. *Id.* at 1032.

64. 463 U.S. 880 (1983).

65. Tex. Code Crim. Proc. Ann. art. 37.071(b)(2) (Vernon 1981).

66. *Barefoot,* 463 U.S. at 896–97.

67. *Id.* at 901.

68. *Id.* at 929 (Blackmun, J., dissenting).

69. 429 U.S. 190 (1976).

70. *Id.* at 204.

71. Oliver Wendell Holmes, *The Path of the Law*, 10 Harv. L. Rev. 457, 469 (1897).
72. 481 U.S. 279 (1987).
73. *Id.* at 287.
74. *Godfrey v. Georgia*, 446 U.S. 420, 427 (1980) (emphasis added).
75. *Gregg v. Georgia*, 428 U.S. 153, 200 (1976).
76. John C. Jeffries, *Justice Lewis F. Powell, Jr.: A Biography* (1994).

Chapter 5

1. New York World Telegram and Sun, April 12, 1958.
2. I. Bernard Cohen, *Science and the Founding Fathers: Science in the Political Thought of Jefferson, Franklin, Adams, and Madison* (1995).
3. *See generally, Hearing of the House Commerce, Justice, State and Judiciary Subcomm.*, April 25, 1996 (chaired by Representative Harold Rogers).
4. *Id.*
5. Connie Mack, *Letter to the Editor, Targeting Congressional Excess First*, Washington Post, August 27, 1995, at C8.
6. *House Hearing*, April 25, 1996.
7. *Id.*
8. Keith Bradsher, *Agency Clones, 200 Lose Jobs, As 'Example' Cut by Congress Aims at Credibility*, Commercial App., Oct. 1, 1995, at A18, available in 1995 WL 9365774.
9. *Department of Energy's Superconducting Super Collider Project: Hearing on Bill Before the Senate Subcomm. on Energy Research and Development of the Senate Comm. on Energy and Natural Resources*, 2 S. Hrg. 102–65 (April 16, 1991).
10. *Basic Science Budget and SSC: Hearing on Bill Before the Senate Subcommittee on Energy Research and Development of the Senate Committee on Energy and Natural Resources*, 2 S. Hrg. 100–730 (April 12, 1988).
11. *See The Superconducting Super Collider Project: Hearing on Bill Before the House Committee on Science, Space, and Technology*, at 19, May 26, 1993.
12. Outlook, Washington Post, Sunday, October 21, 1990.
13. 139 Cong. Rec. S12674-03 (1993).
14. *Id.*
15. *Importance and Status of the Super Conducting Super Collider: Joint Hearing Before the Senate Committee on Energy and Natural Resources and the Senate Subcommittee on Energy and Water Development of the Committee on Appropriations*, S. Hrg. 102–938, June 30, 1992.

16. *Senate Hearing*, April 16, 1991, at 35.

17. *Superconducting Super Collider: Joint Hearing Before the Senate Committee on Energy and Natural Resources and the Senate Subcommittee on Energy and Water Development of the Committee on Appropriations*, 66 S. Hrg. 103–185 (August 4, 1993).

18. *Id.* at 67.

19. *Senate Hearing*, April 16, 1991, at 11.

20. *Id.*

21. Lee Smolin, *The Life of the Cosmos* (1997).

22. Michael D. Lemonick, *The $2 Billion Hole (Supercollider)*, Time, November 1, 1993, at 69.

23. Steven Weinberg, *Before the Big Bang*, The New York Review of Books 16 (1997), reviewing Martin Rees, *Before the Beginning: Our Universe and Others* (1997).

24. Robert L. Park, Opinion/Editorial, *Shelving the Star Trek Myth*, New York Times, July 12, 1997, at 19.

25. 142 Cong. Rec. S9804-01, *S9811 (1996).

26. *Id.*

27. Oscar Wilde, *Lady Windermere's Fan*, Act 3 (1892).

28. Mary Shelley, *Frankenstein; or, The Modern Prometheus*, ch. 5 (1818).

29. *Id.*, letter 4.

30. *Scientific Discoveries in Cloning: Challenges for Public Policy: Hearing Before the Senate Subcomm. on Public Health and Safety of Senate Comm. on Labor and Human Resources*, 1997 WL 8219967 (March 12, 1997); statement of Dr. Ian Wilmut.

31. *Cloning—Challenges for Public Policy: Hearing Before the Senate Subcomm. on Public Health and Safety of the Senate Comm. on Labor and Human Resources*, 1997 WL 8220364 (March 12, 1997); statement of Senator Bond.

32. 143 Cong. Rec. H713-02 (1997).

33. *Senate Hearing*, March 12, 1997.

34. Lance Morrow, *When One Body Can Save Another*, Time, June 17, 1991, at 54.

35. Ogden Nash, *Hard Lines* (1931). The full passage is
    Candy
    Is dandy
    But liquor
    Is quicker.

36. These numbers refer only to the second-generation rats studied in the Canadian research. One of the first-generation control group rats did develop bladder cancer out of the 74 in the study. Only seven of the first-generation experimental rats developed cancer, a result that was not statistically significant (at the $p < .05$ level).

37. *Proposed Saccharin Ban—Oversight: Hearings Before the House Subcomm. on Health and the Environment of the House Comm. on Interstate and Foreign Commerce,* at 41 (March 21 and 22, 1977).

38. *Id.*

39. *Id.* at 84.

40. *The Banning of Saccharin, 1977: Hearing Before the Senate Subcomm. on Health and Scientific Research of the Senate Comm. on Human Resources,* at 6 (June 7, 1977).

41. *House Hearing,* March 21 and 22, 1977, at 248 (testimony of Dr. Sidney Wolfe, Health Research Group).

42. *Id.* at 17.

43. *Cancer Testing Technology and Saccharin,* U.S. Congress, Report of the Office of Technology Assessment (1977).

44. Senate Hearing, June 7, 1977, at 2.

45. *Id.*

46. Federal Food, Drug, and Cosmetic Act, Pub. L. 105–22, 21 U. S. C.A. § 343(o)(1).

47. H.L. Mencken, *A Little Book in C Major,* 19 (1916).

48. Winston Churchill, *Speech,* Hansard, col. 206 (November 11, 1947).

Chapter 6

1. This chapter, by necessity, can be only a cursory overview and sampling of agency practice with the intention of learning more about an important intersection of law and science. It is not, by any means, a comprehensive study of bureaucracy. For an excellent study of organizational practices of government agencies, *see* James Q. Wilson, *Bureaucracy: What Government Agencies Do and Why They Do It* (1989).

2. Lisa Seachrist, *NBAC Down to Wire: Reports Indicate Ethics Panel Favors Legislation,* 8 Bioworld Today 108 (June 5, 1997).

3. Ronald Bailey, *The Twin Paradox: What Exactly Is Wrong with Cloning People,* 29 Reason 1 (May 1, 1997).

4. This section on sport utility vehicles is based on several sources, including Keith Bradsher, *A Deadly Highway Mismatch Ignored: Light Trucks,*

Heavy Risk, NewYork Times, September 24, 1997, at A1, C6; *Overview of Vehicle Compatibility/LTV Issues*, NHTSA Report (February, 1998), available at http://www.nhtsa.gov8o/cars/problems/studies/LTV; *The Aggressivity of Light Trucks and Vans in Traffic Crashes*, NHTSA Report #980908, available at http://www-nrd.nhtsa.dot.gov/nrd1o/aggressivity/documents/98098/98098.htm.

5. Bradsher, note 275, at C6.

6. Rudyard Kipling, *The Second Jungle Book*, *The Law of the Jungle*, st.1 (1895).

7. For excellent histories of the wolf in the United States, *see* Hank Fisher, *Wolf Wars* (1995), and Barry Lopez, *Of Wolves and Men* (1978).

8. 16 U.S.C. §§ 1531–44 (1994).

9. *See generally* Thomas McNamee, *The Return of the Wolf to Yellowstone* (1997).

10. The act can be found at 16 U.S.C. §§ 1531–44 (1994).

11. 50 C.F.R. § 17.11 (1994).

12. 16 U.S.C. § 1531(a)(3) (1994).

13. Luke 10:3.

14. McNamee, note 280, at 34.

15. *Restoration of Gray Wolves to Yellowstone National Park: Hearing Before the House Subcommittee on National Parks and Public Lands*, Serial No. 101-89, at 3–7 (July 20, 1989). Although Owens was technically a "freshman" in 1986, he was no stranger to the House of Representatives. He had served from 1972–1974 before giving up his seat to run for the Senate. He lost that effort, and also subsequently lost a bid for governor in Utah in 1984. Owens again gave up his House seat for a failed Senate bid in 1992.

16. *Northern Rocky Mountain Gray Wolf Restoration Act of 1990: Hearing Before the Senate Subcommittee on Public Lands, National Parks and Forests*, S. Hrg. 101-983, at 3–5 (September 19, 1990).

17. *Wolves in Yellowstone Park: Hearing Before the Senate Subcommittee on Parks, Historic Preservation, and Recreation*, S. Hrg. 104-70, at 3 (May 23, 1995).

18. 16 U.S.C. 1539(j) (Section 10(j) of the act).

19. *House Hearing*, December 11, 1987.

20. *Senate Hearing*, May 23, 1995, at 3.

21. *House Hearing*, January 26, 1995, at 28.

22. *House Hearing*, July 20, 1989, at 13.

23. *House Hearing*, July 20, 1989, at 21.

24. *Id.* at 8.

25. *Id.*

26. *General Accounting Office: A Controversial Issue Needing Resolution*, 52 (1979).

27. *Senate Hearing*, September 19, 1990, at 78.

28. Charles Dickens, *Nicholas Nickleby* (1839).

29. *See House Hearing*, July 20, 1989, at 155 (testimony of Marion Scott).

30. Ken Miller, *Is Return of the Wolf a Social Issue or an Economic Attack?* Great Falls Tribune, October 18, 1993, at A6 (comment of rancher John Matovich).

31. *House Hearing*, January 26, 1995, at 19.

32. *See Reintroduction and Management of Wolves in Yellowstone National Park and the Central Wilderness Area, A Report to the United States Congress by the Wolf Management Committee* (1991).

33. *House Hearing*, January 26, 1995, at 2.

34. *Id.* at 23.

35. Tom Kenworthy, *Babbitt Finds Relocation Program Has Hill's Wolves Growling at Him*, Washington Post, January 27, 1995.

36. House Hearing, January 31, 1996.

37. *See* 16 U.S.C. § 1539(j)(2)(A).

38. Bill Loftus, *Should the Wolf Return? Hearings Around the State Today Aimed at Gauging Public Feeling Towards Reintroduction*, Lewiston Morning Tribune, August 31, 1993, at 8A.

39. Scott Armstrong, *Plan to Release Wolves in Yellowstone Park Creates Controversy*, Christian Science Monitor, August 24, 1993, at 10.

40. Tom Kenworthy, *Sightings Raise Possibility That Wolves Are Outrunning Red Tape*, Washington Post, October. 18, 1992, at A3.

41. *House Hearing*, January 26, 1995.

42. Aldo Leopold (1949), *quoted in* L. David Mech, *At Home with the Artic Wolf*, 171 National Geographic, 562, 565 (May, 1987).

43. Karen Brandon, *9,000-Year-Old Skeleton Spurs Legal Battle; Indians Seek to Deny June Analysis of Possible Caucasian*, Chicago Tribune, August 31, 1997, at 3.

44. This section is generally based on a variety of sources, including *Fact Sheet: Health and Environmental Effects of Particulate Matter*, United States Environmental Protection Agency Office of Air & Radiation, Office of Air Quality Planning Standards, July 17, 1997, available at

http://134.67.104.12/naaqsfin/pmhealth.htm; *Fact Sheet: EPA's Revised Ozone Standard*, United States Environmental Protection Agency, Office of Air & Radiation, Office of Air Quality Planning & Standards, July 17, 1997, available at http://134.67.104.12/naaqsfin/o3fact.htm; *Final Revisions to the Ozone and Particulate Matter Air Quality Standards: Current and Revised Standards for Ozone and Particulate Matter*, EPA Report, available at http://earth1.epa.gov/oar/oaqps/ozpmbro/current.htm; *National Ambient Air Quality Standards for Particulate Matter*, Final Rule, 40 CFR Part 50 (July 18, 1997); *National Ambient Air Quality Standards for Ozone*, Final Rule, 40 CFR Part 50 (July 18, 1997).

45. *American Lung Assoc. v. Browner*, 884 F.Supp. 345 (D. Ariz. 1994).

46. *Impact of Clean Air Rules on Agriculture: Hearing Before the House Committee on Agriculture* (September 16, 1997).

47. *Testimony of Carol Browner: Hearing Before the House Committee on Agriculture* (September 16, 1997).

48. *Prepared Testimony of George T. Wolff, Chair, EPA's Clean Air Scientific Advisory Committee's Panel on Ozone and PM: Hearing Before the House Subcommittee on Commercial and Administrative Law of the Judiciary Committee* (July 29, 1997).

49. *Id.*

50. 42 U.S.C. 7408 § 109(b)(1).

51. *National Ambient Air Quality Standards for Particulate Matter*, Final Rule, 40 CFR Part 50 at 3 (1997).

52. Senators Imhofe and Breaux and Representatives Klink and Upton, News Conference, (June 25, 1997).

53. *Memorandum from Bill Clinton to the Administrator of the Environmental Protection Agency*, at 8 (July 16, 1997) (on file with author).

54. ___ F.3d ___, 1999 WL 300618 (D.C. Cir. 1999).

55. *Id.* at *1.

56. *Id.*

57. *Id.* at *4.

58. *See Id.* at *6.

59. Lewis Carroll, *Alice's Adventures in Wonderland*, st. 18 (1865).

60. *American Trucking Associations*, 1999 WL 300618 at *7.

Chapter 7

1. Karl Popper, *Conjections and Refutations* (1962).

2. *Reauthorizing the Endangered Species Act: Hearing Before the Senate*

Subcomm. on Clean Water, Fisheries and Wildlife, 1994 WL 266928 (June 15, 1994), opening statement of Senator Bob Graham.

3. *Id.* (emphasis added).

4. *Reauthorizing the Endangered Species Act: Hearing Before the Senate Subcomm. on Clean Water, Fisheries and Wildlife*, 1994 WL 14188967 (June 15, 1994), statement of Edward O. Wilson.

5. *Id.*

6. *Id.*

7. *Reauthorizing the Endangered Species Act: Hearing Before the Senate Subcomm. on Clean Water, Fisheries and Wildlife*, 1994 WL 530640 (September 29, 1994), statement of Senator Bob Packwood.

8. *Id.*

9. *Ulmer v. State Farm Fire & Casualty Co.*, 897 F. Supp. 299 (W.D.La. 1995).

10. *Id.* at 303.

11. These reference books give lawyers, judges and policy makers detailed guidance on a wide range of scientific topics. *See Scientific Evidence Preference Manual* (Federal Judicial Center, 1994); David L. Faigman, David H. Kaye, Michael J. Saks & Joseph Sanders, *Modern Scientific Evidence: The Law and Science of Expert Testimony* (West Pub. Co. 1997 & 1999 Supplement).

12. Plato, *Phaedo* (ca. 30 B.C.).

INDEX

◆

Frankfurter, Felix, 95
Franklin, Benjamin, 123
Franklin, George, 59, 60, 86
Fraser, Persifor, 3
Freedom of speech, 114
Free Exercise Clause, 19, 202
Free will, vs. biologic determinism, 27
Freud, Sigmund, 88
Frye v. *United States*, 62–63

Galileo Galilei, 12, 14, 16–18, 19, 105
Galton, Francis, 10–11
Gardner, Sherwin, 147
Gatekeepers, judges as, 60–64, 77, 79–82, 86–88, 195–197
General Electric Co. v. *Joiner*, xii-xiii
Geocentric theory of universe, 14
Ginsburg, Ruth Bader, 30, 96
Glenn, John, 139, 140
Global warming, 174
Goethe, Johann Wolfgang von, ix
Good, Dorcas, 35
Good, Sarah, 34
Graham, Bob, 192
Gramm, Phil, 132–133
Gravity, 130
Greenhouse, Linda, 114
Greenhouse effect, 174
"Greenhouse" effect, 114
Griffin, Jane, 126–127
Griggs, William, 34
Grigson, Dr., 111
Griswold v. *Connecticut*, 119
Gunther, Gerald, 96

Haag, Ernest van den, 102
Hamilton, Alexander, 113
Hand, Learned, 97
Handwriting identification, 2–5, 77–78, 86, 202–203
Hansen, James, 164, 172
Harlan, John, 98

Hatch, Orrin, 48, 49
Hawking, Stephen, xi, 13
Hayakawa, Senator, 148
Heckman, Nicholas, 2
Heliocentric theory of universe, 16–18, 19, 20
Hendricks, Leroy, 27–32
Higgs, Peter, 130
Higgs particle, 130, 131
Hinkley, John, 28–29
Hobbs, Deliverance, 35
Holmes, Oliver Wendell, 66, 99, 116
Holstrom, Linda, 71
Holtzmann, Steve, 157
House members. *See* Legislators
Hubbard, Elizabeth, 34
Hubble telescope, 140
Human cloning, 141–145
Huxley, Aldous, 201
Huxley, Thomas, 14, 76
Hyde, Henry J., 48, 49

Innumerates Anonymous, 198–199
Insanity defense, 27–32
Insanity Defense Reform Act, 29
Interpretive fact concept, 104–108, 110, 118–121
Irresistible impulse, insanity defense and, 28

Jacobs, Andrew, Jr., 148
Jarvis, William, 93
Jefferson, Thomas, 11, 91–93, 123
John Paul II, 19, 143
Johnson, Pamela, 66
Johnston, Bennett, 129, 132, 134, 135
Johst, Hanns, 13
Jones, Robert, 200
Judges. *See also* Supreme Court
    assessment of scientific evidence by, 60–64, 77, 79–82, 86–89, 195–201
    scientific ignorance of, 88–89, 110–113, 115–121

Judges (*continued*).
  technical information sources for, 200
Judicial review. *See also* Constitutional interpretation
  establishment of, 91–93
Jurek v. *Texas*, 111
Juries
  assessment of scientific evidence by, 64–66
  scientific ignorance of, 53–55
  size requirements for, 108–110

Kaczynski, Theodore, 85
Kafka, Franz, 59–60
Kansas Sexual Predator Act, 29
Kansas v. *Hendricks*, 30–32
Kaye, David, 109
Kennedy, Anthony M., 31, 96
Kennewick Man, 176–178
King, Ray, 40
Klaas, Polly, x
Klink, Ron, 180
Kozinski, Judge, 63
Kuhn, Thomas, 56
Kumho Tire Co. v. *Carmichael*, xiii, 79–82, 86
Kunstler, William, 70

Law
  religion and, 8–9, 33–38, 201–202
  science and. *See* Science and law
Law schools, curricular focus in, xii
Lawyers, scientific ignorance of, 53–55, 197–201
Lederman, Leon, 13, 137
Legal Tender Cases, 95
Legislative hearings, 125
Legislative process, science in, 122–152
Legislators. *See also* Congress
  scientific understanding of, 50–57, 123–126, 197–204

technical information sources for, 126–127, 199–200
Leibeck, Stella, 65
LeJeune, Jerome, 43
Leopold, Aldus, 171
Leo X, 15
Lethal Drug Abuse Prevention Act of 1998, 49–50
Lie detector tests, 84, 196–197
Little, Judge, 197
Lobbying
  administrative rulings and, 158–160
  legislation and, 123–124
Lochner v. *New York*, 98–99, 100, 113
Lung cancer, smoking and, 51, 175–176
Luther, Martin, 16

Mack, Connie, 126
Madison, James, 123, 124
Marbury v. *Madison*, 92–93, 97, 113, 118
Marlenee, Representative, 164
Marlow, Christopher, ix
Marshall, John, 91–93, 101
Marshall, Thurgood, 101
Martin, Luther, 95
Mass toxin torts, 65, 192, 193
Mather, Cotton, 36, 37
Mather, Increase, 36, 37
McCleskey v. *Kemp*, 116–118
McClure, James, 162–163
McKenna, Judge, 77–78, 202
Medical treatment, refusal of, 45–50
Mencken, H. L., 20, 43, 151
Mental illness, criminal culpability and, 27–32
Merrill, Richard, 147
Metaphysics, 12–13
Meteors, 174
M'Naghten, Daniel, 28
M'Naghten test, 28–29
Molineux, Roland, 1–5, 6

Powell, Lewis, 110, 116, 117, 118

Precedent, 95, 111–112

Prediction of violence, 31–32, 69–70

Preembryos, legal status of, 39–45

Preponderance of the evidence
standard, 66–70

Presidency, as nondemocratic
institution, 47–48

President, agency supervision by,
154–155

Pressure groups
administrative rulings and, 158–160
legislation and, 123–124

Privacy rights, 119

Private Adjudication Center, 200

Pseudoscience, 70–76
evaluation of, 86–87

Psychiatric illness, criminal
culpability and, 27–32

Ptolemy, 15

Putnam, Ann, 34, 35

p value, 67, 68, 69

Qwerty effect, 83

Racial segregation, 102–106

Radiocarbon dating, 178

Rape shield statutes, 75

Rape trauma syndrome, 71, 75–76, 84

Rawls, John, 96

Reagan, Ronald, 162

Rehnquist, William H., 31, 64

Religion
Constitutional protections and
restrictions of, 18–19, 201–202
establishment of, 19
law and, 8–9, 33–38, 201–202
science and, 8–9, 11–26

Reno, Janet, 49

Representatives. *See* Legislators

Repressed memories, 37, 59, 86

Reproductive technology, 39–45

Research. *See* Scientific research

Right to die, 45–50

Robinson, Pat, 19

Robinson, William J., 11

Roe v. *Wade*, 42, 105–108, 118, 119,
120–121

Roosevelt, Franklin D., 9, 153

Saccharin, bladder cancer and,
145–150

Safety regulations, automotive,
158–160

Salam, Abdus, 130

Salem witch trials, 33–37

Savoie, Robert, 196–197

Sayres, Gilbert B., 3

Scalia, Antonin, 24–26, 31, 96

Scanlon, Thomas, 96

School, teaching of evolution in,
19–26

Schweiker, Richard, 149

Science
administrative agencies and,
172–179
big-ticket, 128–141
big vs. small, 133–137
early concepts of, 7
forensic, 76–82
judges's understanding of, 110–113,
115–121, 197–204
legal definition of, 51
in legislative process, 122–152
legislators' understanding of,
50–57, 123–126, 197–204
market theory of, 82–86
objective of, 7–8
as organized common sense, 14–15
policy aspects of, 70–76
postmodernism and, 177–178
privileged status of, 176
religion and, 8–9, 11–26

Science and law
agenda-setting and, 110
cultural conflicts between, 56–57